SCIENCE, CELLS AND SOULS

Other books by Neville Moray

Cybernetics: Machines with Intelligence

Introduction to Psychology for Schools

Attention: Selective Processes in Vision and Hearing

Listening and Attention

Mental Workload: Theory and Measurement

Human Factors Research and Nuclear Safety (with B. Huey)

Human Error: Cause, Prediction and Reduction (with J.W.Senders)

Robotics, Control and Society (with W.R.Ferrell and W.B.Rouse)

Ergonomics: Major Writings

SCIENCE, CELLS AND SOULS

AN INTRODUCTION TO HUMAN NATURE

Neville Moray

authorHOUSE®

AuthorHouse™ UK
1663 Liberty Drive
Bloomington, IN 47403 USA
www.authorhouse.co.uk
Phone: 0800.197.4150

© *2014 Neville Moray. All rights reserved.*

No part of this book may be reproduced, stored in a retrieval system, or transmitted by any means without the written permission of the author.

Published by AuthorHouse   12/05/2014

ISBN: 978-1-4969-9699-2 (sc)
ISBN: 978-1-4969-9698-5 (hc)
ISBN: 978-1-4969-9673-2 (e)

Any people depicted in stock imagery provided by Thinkstock are models, and such images are being used for illustrative purposes only. Certain stock imagery © Thinkstock.

This book is printed on acid-free paper.

Because of the dynamic nature of the Internet, any web addresses or links contained in this book may have changed since publication and may no longer be valid. The views expressed in this work are solely those of the author and do not necessarily reflect the views of the publisher, and the publisher hereby disclaims any responsibility for them.

For my wife

Angela angelusque

What a piece of work is a man!
How noble in reason! How infinite in faculty!
In form and moving how express and admirable!
In action how like an angel!
In apprehension how like a god!
The beauty of the world! The paragon of animals!
And yet, to me, what is this quintessence of dust?

                    Shakespeare: Hamlet: Act 2, Scene 2

Human beings want to understand themselves, and in our time such understanding is pursued on a wide front by the biological, psychological, and social sciences. One of the questions presented by these forms of self-understanding is how to connect them with the actual lives all of us continue to lead, using the faculties and engaging in the activities and relations that are described by scientific theories.

                Thomas Nagel. 2012. New York Review of Books.

Everyone has a theory of human nature.

                        Steven Pinker, *The Blank Slate*. 2002.

Table of Contents

Foreword xiii
Acknowledgements xv

Part 1   Thinking About Stories
              Chapter 1. Introduction 3

Part 2   Thinking About Thinking
              Chapter 2. Causes and Explanations 21
              Chapter 3. Ways of Thinking: Philosophy 35
              Chapter 4. Ways of Thinking: Science 56
              Chapter 5. Probability 79

Part 3   Thinking About Humans
              Chapter 6. Names, Nouns and Things 101
              Chapter 7. Life 113
              Chapter 8. Evolution 136
              Chapter 9. Making a Person 153
              Chapter 10. Abilities: Nature and Nurture 167
              Chapter 11. Artificial Intelligence: Sharing
                              the World With Artefacts 184
              Chapter 12. Brain, Mind and Consciousness 209
              Chapter 13. Free Will and Responsibility 236
              Chapter 14. All About Souls 257

Part 4   Thinking It Over
              Chapter 15. Epilogue 281

Appendix 1 285
Appendix 2 287
Bibliography 289

References and URLs shown in footnotes will also be found in the Bibliography.

List of Illustrations

Figure 1.1   Why did Salterella jump
Figure 2.1   What caused Salterella to jump?
Figure 5.1   The Normal Distribution (bell curve)
Figure 5.2   A discrete probability distribution
Figure 7.1   A schematic representation of an eukaryotic cell
Figure 8.1   *Copilia quadrata*
Figure 8.2   Species of copepods showing evolution of vision
Figure 8.3   Goats climbing trees in Morocco
Figure 9.1   Three ways to represent the glucose molecule
Figure 9.2   The relation of Sucrose, Fructose and Glucose
Figure 9.3   Haemoglobin and Chlorophyll
Figure 9.4   The anatomy of a typical neuron
Figure 9.5   The Krebs Cycle
Figure 10.1 The properties of the Normal Curve
Figure 10.2 Overlapping curves of IQ distributions
Figure 11.1 A prosthetic hand
Figure 11.2 A Pitts-McCulloch neuron
Figure 13.1 A neural information processing chart
Figure 14.1 The development of a human

All figures are original or in the public domain except where specifically acknowledged.

List of Tables

Table 11.1   Truth table for an AND function
Table 11.2   Truth table of a Pitts-McCulloch neuron
Table 11.3   Truth table for an AND function expressed as multiplication in binary arithmetic.

Foreword

This book owes its origins to a modern anachronism, the Men's Book Club of the Riviera. Its author, Neville Moray, is an anachronism too. He is a distinguished experimental psychologist and human factors engineer, well known to psychologists for his early work in selective attention, and more widely, for his technical work in mental workload measurement and human-machine interaction. It is not uncommon to find an author prepared to accept the challenge of explaining human nature in terms the lay person can understand. But to find one who in doing so can range from principles of Aristotelian logic to an understanding of machine intelligence and how living things are organised and have evolved, is rare. Polymaths are supposed to be extinct.

It is a long time since I read a book of anything like comparable originality and breadth. Moray writes for the lay person, and addresses some well-worn questions, such as the nature of consciousness, or whether free will exists, but his take is refreshingly different from any other I have read. He explains the basis of science in some detail, and his insistence on the indivisibility of scientific knowledge is relentless. There are no 'ghosts in the machine' here, no 'feather beds for falling Christians', no concessions to Cartesian dualism. Yet there is no comfort either for strict materialists. On the contrary, this is a writer not afraid to call a soul a soul, though I suspect that what he means by 'soul' will not be quite what you may expect. His book has the explicit aim of raising questions in the minds of readers, about how we should conceptualise having a soul, or a will in the context of neuroscience, or how we should explain what it means to speak of human nature.

Neville Moray

Despite the author's intentions, it is not always an easy read, and it challenges all the time. I did not agree with everything I read, but I enjoyed reading it. I kept going back for more. It certainly made me think, and I am very inclined to turn my students loose on it.

John Elliott, Associate Professor,
Department of Psychology,
University of Singapore

Acknowledgements

First I have to thank The Men's Book Club of the Riviera. I have been a member of that group for ten years, and we have had many discussions that were stimulated by contemporary books on computers, biology, history, philosophy, law and other topics that raised questions about human nature. In the end so many ideas arose from those discussions that I decided to try to summarize them in a book. This is the result, the thoughts of someone who grew up originally with a religious background which he has now abandoned, and whose profession involved psychology, philosophy, computing and science. Of course my fellow members of the Book Club cannot be held responsible for the content of the book, but they are responsible for its existence. I hope they will accept it as showing how much I value their friendship.

I owe an enormous intellectual debt to Sir Anthony Kenny. We first met 60 years ago when we were both graduates at Oxford, and he guided me in my first explorations of philosophy. I have followed his work since then with the greatest admiration both for his abilities as a philosopher and as a teacher. He is not however to be held responsible for any egregious errors you may find in the book. Another influence on my philosophical thinking over many years is Professor William Ruddick, of New York University. The late Timothy Firth has a unique role in that like me he began life as a Catholic, like me gave up that belief after many years, and helped and encouraged me to explore alternative visions of the world while trying to retain what was of intellectual and human value from our earlier tradition. Professor John Frisby of the University of Sheffield has been a long-time collaborator in discussions of psychology and modelling of the nervous system. John Elliott, Associate Professor of the University of Singapore gave critical advice and editorial help far above and beyond the call of friendship. Professor Ray Nickerson of Tufts University, Professor Tom Sheridan of MIT and Professor Peter Hancock of the University of Central Florida having allowed me to comment on early drafts of books they have written have in turn helped me by criticising my manuscript, as well as being colleagues in the field of engineering psychology and ergonomics for many years. I must thank also two friends since student days, Professor George Sitwell and Dr. Penelope Rowlatt, for their encouragement and intellectual stimulation. As always on these occasions, the faults that remain are mine.

Finally I must thank my wife Angela for her criticism, editing and proof-reading, and for being a constant source of encouragement: what I owe her is I hope expressed in the dedication.

PART 1
THINKING ABOUT STORIES

Chapter 1

Introduction

"...difficulty in making my own thoughts sufficiently distinct and clear to communicate them, . . . in writing. They are mature enough to climb up and chirp on the edge of their Birth-nest; but not fledged enough to fly away, tho' it were but to perch on the next branch."

Samuel Taylor Coleridge. 1826

It is not comfortable in the nest. Today many ideas which we took for granted about human nature are not left to mature in peace. Science and technology seem like cuckoos that may at any moment force us out of the nest and leave us gasping on the ground. It would be better, certainly, to be brave enough to make our first flight voluntarily. For whom then is this book intended? To use Coleridge's metaphor it is for those ready to leave the comfort of the intellectual nest in which they have matured. It is an introduction to ideas for scientists who are ill at ease with the philosophical implications of their discoveries, and for people who don't understand what is happening in scientific research, for those interested in the intellectual excitement of our times. It is an introduction, not an encyclopaedic treatment.

Although we are all human, human nature is full of mysteries. It is all too easy to opt for a simple approach through science, religion, or personal intuition. Take any group of well-educated and successful people with time to think about their own natures. They may have had successful careers in banking, commerce, entertainment, journalism, industry, law or even the sciences. They may be widely read and interested in the modern world. Some will be sympathetic to religion, the majority probably not. But two things that are common is that they often find it difficult to understand and evaluate scientific discoveries, and usually have not looked in detail at the ideas, the

Everyday Stories as we shall call them, that they take for granted about human nature. Those ideas, whether asserted or denied, include words like soul, mind, consciousness, self and will.

Such ideas used to be a matter for philosophy. People used to speculate endlessly about, "What is life?" "Does our soul survive death?" "Do we have free will?" It was enough just to sit and think about them. But today such questions seem increasingly a matter for science. Can we make life? Can education improve intelligence? What would it be like to clone a human? Does the electrical activity of the brain mean that free will is an illusion? Will we soon share the world with intelligent machines? If one has not studied science it is difficult to assess modern research: if one is unfamiliar with philosophy it is difficult to criticize one's intuitions.

This is a good time to study human nature. We understand the chemistry of the stars, the age and origin of the universe, and the fundamental structure of matter. While the twentieth century was the age of physics, we are now in the age of biology. We understand ever more about the biochemistry of living bodies, the basis of heredity, the electrical activity of the brain and the physiology of nerve cells. We have invented devices that mimic muscles, sense organs and human thought. Some claim that we can tell from patterns of brain activity when people are telling the truth, when they make voluntary choices, and even when their religious belief is active.

What are we to make of such claims, which seem to be about mental or spiritual activities, but are measured by physical events? How does the new science relate to the classical ideas about our nature that have come down to us as "the wisdom of the ancients", the subject matter of philosophy? The answers lie on the border of biology, philosophy, science and technology, a border that is getting rather crowded these days. The underlying theme of this book is that mere scientific knowledge does not by itself deepen our understanding of human nature. We must relate it to ideas that were common before the rise of science. But neither is philosophical speculation enough: we must examine different kinds of knowledge, how they relate one to another, and see how a synthesis can help us better to understand what we are.

There are traditional words that seem intrinsic to human nature. I call them *Fundamental Words* and they include "self", "body", "mind", "will", "soul", and "life". Are such words the names of parts of a person, non-physical, immaterial, or "spiritual"? People are loath to abandon them, but do they fit in with the new scientific discoveries? Many books today discuss modern neuroscience and biology as if their success means there is ever less reason to retain older ways of talking. That risks impoverishing our understanding. In this book we will try to connect the old and new ways of thinking. We will go backwards and forwards in the history of thought far beyond what is usually considered in contemporary discussions of, for example, neuroscience. We will concentrate on the nature of the individual human. What we decide about free will not tell us how to treat a murderer, but will help us to understand whether people are responsible for what they do. If people are indeed responsible for their actions that should make a difference to the law, but exactly what difference will depend on a particular culture, and that I leave to others.

I want to examine different kinds of knowledge, how they relate one to another, and how a synthesis can help us better to understand human nature. So we have to start by looking at science and philosophy themselves. Then we can look at the Fundamental Words and what science has to say about them. Don't always expect answers. Indeed I shall have succeeeded only if at the end of the book readers have more questions than they had at the beginning. But at least they will be their own questions.

Although the chapters are not in a strictly logical order, they fall broadly into two sections. It would be nice to plunge directly into what science has discovered about life and the mind, but we can't really do that. If we are to accept the claims of science which are made with such certainty we need to understand something of the nature of science. After all, even Norbert Wiener, who founded cybernetics, maintained that in science there was no such thing as certainty. So we start by thinking about the nature of science as such, and about probability. How can we reason about ideas themselves? How can we relate what we take for granted to challenges posed by new discoveries? The same applies to philosophy. How can we use it to examine how we think?

The choice of topics may seem rather arbitrary. Why is there a chapter on evolution? Why is intelligence discussed at length? Why is there a chapter on prostheses and artificial intelligence, on probability? The reason is that in discussions over many years I have found that people often do not realise what modern science and philosophy say about such topics. People will often argue passionately for positions without really understanding the evidence. For example, they will argue the "nature vs. nurture" issue in relation to intelligence without knowing that almost all the research is based on IQ tests, without knowing how IQ is measured, and without knowing what recent research says about the nature of race and heritability. Many people surprisingly have only the vaguest idea of what is meant by "evolution", and what the evidence is for it. And many, who hold strong opinions about whether machines can think, do not know what has been achieved by artificial intelligence and advances in computing, or what the general theory of machines asserts. I have chosen topics that seem of general concern, but where there seems a need for at least some familiarity with recent advances.

All too often books about modern science take an aggressive stance toward people who believe in older philosophical and religious ideas. In turn, people who still think in terms of souls, wills and minds are dismissive of modern science. What a shame! The richness of human nature requires an equally rich approach. So we will look at recent scientific discoveries but also at classical ways of talking about humans. We will think about *how* to think about humans. Then we will go on to see what new discoveries tell us about genetics, abilities and the brain, and how they change our ideas of life, mind, and even soul. We are looking for a synthesis, not a knock down and drag out defeat of some imagined opponents' views or straw men.

STORIES

Stories are efficient summaries of reality, but that isn't all they are. Stories have an arc, they put constraints on the future - when you've heard the first half there are some things which are more likely in the second, and some less. I'm sure our minds use stories because they describe the way the world is and because they say something about how the world could or will be.

Tom Stafford[1]

The Story of Human Nature

We are all human, so in one sense we start level when we try to understand human nature. People differ from gender to culture, and from the places in which they live to the time at which they were born; but they share common abilities and experiences that are the same down the ages and across environments. What makes today a particularly good time to analyse human nature is that new ways to examine it have appeared particularly in the last 50 years. These give us insights into human nature through the application of biological, behavioural and physical sciences.

Compare what we know today with what we knew just a lifetime ago. A neurologist wrote in the late 1940s,

> It remains sadly true that most of our present understanding of mind would remain as valid and useful, if for all we knew, the cranium were stuffed with cotton wadding.[2]

Today we can use functional magnetic resonance imaging (fMRI) and similar techniques to watch from moment to moment parts of the brain become active in cognitive tasks and in response to emotional stimulation. We have an extremely detailed picture of what happens in the microstructure of a neuron (nerve cell) when an impulse passes along it carrying information from one part of the brain to another. Again, consider what a textbook on how to use a microscope, written in 1940, said about the cell:

---

[1] http://www.40kbooks.com/?p=2176
[2] R.W.Gerard, 1946. The Biological Basis of Imagination. *Scientific Monthly*.

> . .the mitochondria. . .(are). . .said to consist of protein and phospholipids and may be seen during life. They can grow, but their functions are not exactly known. It is suggested that they are centres of chemical activity, possibly connected with enzymes. . . . The chromosomes are even more acidic and dye even more strongly, hence their name. When fully formed they contain nucleic acid as such[3].

As we shall see in Chapter 7 we now know that mitochondria are intimately concerned with making energy available to the cell, and that they originated as bacteria that lived symbiotically with human cells. We can even watch individual molecules going about their metabolic operations in the cell. And can you imagine today an article about chromosomes that does not start by talking about DNA? To take another discipline, before the late 1950s no university in the UK had a commercially manufactured digital computer.[4] Today artificial intelligence is a practical reality and millions of people own not one but several computers in the form of desktops, laptops, tablets and smart phones. We know about the atoms that make up our bodies, and the predictions of Quantum Theory about the underlying physics have never been wrong in 100 years. We know much about the mechanisms of living organisms, whether individual cells such as bacteria or the great masses of cells that make up bodily organs such as the brain or the heart; and we are close to synthesizing life. Neuroscience in particular has made huge strides in recent years.

Some say that given all this new knowledge everything we do and experience can be explained by science. But somehow the traditional questions that were formulated long ago by philosophers seem to remain unanswered. The Fundamental Words retain their fascination. What is life? Do we have free will? Do I have a soul? Science, whether physical, biological or psychological, doesn't quite seem to answer these questions satisfactorily. Why not? Is science not enough to understand human nature?

---

[3] H.A.Peacock, 1940. *Elementary Microtechnique.* London. Edward Arnold.
[4] It was only in the late 1960s that I installed the first digital computer to be used by a UK psychology department, at the University of Sheffield.

This book is a starting point from which readers can set off on their own intellectual flights. While some of the topics look straightforward we need to understand how, historically, they have come to be part of our culture. Also, how does science get its certainty if probability plays such a role in scientific research? How is that certainty justified?

So let's begin.

Steve Rose[5], a biologist, once asked why a frog suddenly jumped. He first explained the jump in terms of the physiology and biochemistry of the frog's muscles, and the physics of the molecules in the muscles. He then suggested that the cause of the jump was not just a muscle twitch, but the presence of a snake from which the frog was trying to escape. Now on one hand the physiological story was true. On the other, so was that about the ecology of behaviour. Are there perhaps many ways of talking about a living entity all of which are in some sense true at the same time? Are only some of them scientific, and others of a different kind? Can we pick and choose among them? If not, when should we choose one rather than another?

Let's start with a similar story, also about jumping, but one that is more interesting to humans. Figure 1.1 shows a woman high-jumper[6]. I will call her *Saltarella*, from the Latin word *saltare*, meaning *to jump*. Let's suppose she is trying to win the high jump in the Olympic Games. We can look at her behaviour and achievements as a framework for understanding her as a typical human, with body, mind and will. How can we explain the events in her life, her biography?

---

[5] S.Rose. 1997. *Lifelines*. Oxford. Oxford University Press
[6] With permission of Getty Images.

Figure 1.1 Why did Saltarella jump?

What are the best answers to questions such as, "What caused Saltarella to jump?" "How did she jump?" "Why did she jump at this time and in this place?" "Why was she so great an athlete?" "How did her thoughts translate into action?"

How can we answer such questions?

Telling Stories

By a *story* I don't mean something necessarily untrue, but a more or less self-contained account of some aspect of life. To say that something is a story is to say that in it language will be to some extent constrained, and the possible meanings of words restricted to a well-defined convention. For example, if I tell you a story where a step-mother has a role, you will have different expectations if you know the story is about the life of a divorce lawyer rather than a fairy story. Stories give us something to hang on to: they give sense to our lives. Each story seems to give us *an* answer, however imperfect and incomplete, to *some* questions. Stories can be true or fictional, old or new, of many different kinds; and it is important to understand that different kinds of stories complement rather than contradict one another. Perhaps if we put them all together we will approach the truth. Let's look at some kinds of stories that we might tell about being human.

Everyday Stories

Everyday Stories are stories that the majority of people in a culture believe to be true, and indeed take for granted. Everyday Stories are rarely questioned. "Well, everyone just *knows* that *that* is true, don't they?" An example in our culture is for many people, the "fact" that capital punishment prevents murders. Another is that your horoscope, and the positions of the stars, have a big influence on what happens in your life. If I ask someone why they believe these ideas are true, they tend to reply as I suggested that, "Everyone knows that. . . ", but don't give a well reasoned justification for their belief. In our culture, for many people, even the idea of God is an Everyday Story.

Authoritative Stories

There are some stories about human nature that get their strength from the authority of the storyteller, the person who vouches for the truth of the story. Typically we get such Stories from parents, teachers, books, newspapers or television programmes. We may hear of people, alive or dead, who are so charismatic that they convince us even if we can't fully understand them. The lives and opinions of people such as Albert Einstein, Nelson Mandela, or Florence Nightingale are among such Stories. Those who were called "saints" in the past had this kind of power. The Founding Fathers of the American Constitution told Authoritative stories in the preamble to the Constitution: one *Authoritative* story in their culture says that despite appearances to the contrary all men are born *"equal"*. Another is the Everyday story that people consist of both material and spiritual parts and when people die the spiritual part continues to exist. (This story, which many initially believe on the authority of parents or teachers, is often a Religious story, but need not be. The Greek philosopher Plato held such a view in a non-religious context.) *Historical, Political* and even *Religious* stories are often also *Authoritative* stories.

Philosophical Stories.

Philosophical Stories examine the logic of what we say. Can we be sure that it is not self-contradictory? If we say that some things in a story are *true*, what does that mean? What does it mean to say that one thing *causes* another? If we believe we have free will, what is a will, and how does it work? How can "mental" events occur in the material body?

Philosophy has many meanings. In this book it is treated as an intellectual discipline that applies rigorous analysis to ideas expressed in language. Philosophy examines knowledge we already have and brings out its implications: it does not discover new knowledge, although sometimes philosophical analysis may lead to surprising and completely unforeseen conclusions[7]. Unlike science, which discovers new knowledge, philosophy is concerned with knowledge available in principle to everyone, not just to experts[8].

Scientific Stories.

Many modern stories about human beings are based on science. Science uses empirical, often experimental, methods to investigate our universe (including ourselves) according to strict methodological rules. Scientific stories are among the most powerful Stories that we have, and the major way in which our culture has been shaped in the modern era. Science adds new knowledge that may be very difficult even for experts to obtain and understand.

Most scientific Stories are quite recent, less than 300 years old, and many have been first told in our own lifetime. They include stories about physics, chemistry, biology, psychology and other sciences.

---

[7] An example that might be thought of as discovering "new knowledge" would be Godel's Theorem about mathematics. But the result is implicit in the original knowledge of mathematics, although difficult to deduce.

[8] It is said that Professor Isaiah Berlin once overheard two women on a bus discussing how sad their lives were. One said to the other, "You've got to be philosophical, dear. Don't think about it too much." That is *not* what I mean by doing philosophy!

For example the scientific stories of astronomy and cosmology tell us about the nature of distant parts of the universe, and about its origins. We expect physics to tell us about the nature of matter and the properties of materials and subatomic particles. Biological stories tell us about life and reproduction, how living creatures appeared and how one kind of creature changes into another. Some kinds of psychology tell us the about laws of behaviour, and are backed by stories about how our nerve cells work and what part of the brain does what. (Other kinds of psychology, such as psychoanalysis, do not use scientific methods and are not science.) Applied science tells us how to make artificial devices with desired properties, and lets us design machines that mimic animal and human abilities (the fields of robotics and artificial intelligence). As we shall see, science has particularly effective ways to check whether its Stories are true.

Historical Stories

Historical stories are about events that occurred in a more or less distant past. In some respects they resemble scientific stories, but they differ from the latter because we cannot perform empirical research like that supporting scientific stories. If I doubt that water boils at 100°C, I can heat it again and measure the temperature at which it boils: but if I assert that Napoleon took the road below my house during his return to France from Elba, I cannot watch him a second time and see which road he takes. I may be able to find evidence, including documents, letters, etc., relevant to the story I want to tell, but I cannot "do it again". Historical stories let us try to understand past events, and sometimes to see how events that happened long ago shaped our own lives and times. (Some kinds of science seem to resemble history, particularly some cosmology and some research on the sequence of events in evolution, but the resemblance is superficial.)

Religious Stories

Science and religion need not be mutually antagonistic. Someone who believes Religious Stories can see science as giving an extra richness to their view of how God orders the universe, which was certainly how the great Isaac Newton saw things; while an atheistic scientist will tell a story that does not include a religious chapter. It does not help to be simplistic. If I say "I believe in God" I may mean any of the following:

1. I believe that God exists.
2. I believe that there is a being who cares for me and can be trusted to protect me.
3. I have faith in God in the sense that I am ready to commit myself to a particular way of life, caring for the poor, worshipping God, and so on.

None of these necessarily conflict with science or philosophy. (1) has no direct impact, until the content of my belief is made more explicit. (3) clearly does not have any impact on my approach to scientific or philosophical truth unless my religious belief commands me to do something unscientific. But then what does "unscientific" mean? (2) would only cause conflict if I believed that God's protection was a better way of keeping people from harm than ways proposed by science. Obvious examples are a refusal to vaccinate children or to help people to die who are suffering. But the conflict here is not exactly with science as such. Science may prove that vaccination will protect a person from smallpox or polio. But the decision that it is good to do everything that we can to prevent disease from spreading is not a scientific statement – it is a moral, political or ethical statement; and statements about what is right and wrong are not strictly speaking scientific statements. The belief that one should always apply what scientists currently say about the universe is a philosophical, religious or political assertion, perhaps an Authoritative Story, not in itself a Scientific Story.

If someone denies an isolated scientific fact in the name of religion, what follows? We should ask whether the religious person wants to deny not just *a* particular fact but *all* the facts of science, for we shall see later that science is a set of interlinked claims, so that the science that says that evolution occurs is indirectly linked to the science that makes a mobile phone work. On the other hand we should ask a scientist to recognize when he is proposing a moral or political programme, not a scientific fact, and to justify his moral claim. And we have to admit that there are ways of becoming convinced of the truth or importance of Religious Stories that perhaps are immune from rational discourse. The actor David Suchet has spoken movingly about his experience of reading an epistle of St. Paul and instantly feeling that the life described therein was such that he felt compelled to adopt it. That kind of certainty is like the experience of suddenly falling in love. One's friends may say, "But how can you possibly have fallen in love with her? Don't you know what she is like?" and to the reply, "You just don't see her as I do: to me she is wonderful." Against that kind of experiential certainty the arguments of both science and philosophy are almost irrelevant.

Fantasy and Fiction.

These are Stories that imagine how the world might be or could be, or present a writer's personal view of the world. The only kind of fantasy and fiction which you will find in this book are what I call *fables*, fictional stories that I use to make a point more understandable, a kind of drawn-out metaphor. I will not treat either religion or science as fantasy or fiction: I do not want to spark aggressive disagreements.

Other Stories.

There are many, many kinds of stories, but the Stories above are those with which this book will mainly be concerned. We will have little to say, for example, about political or economic stories in this book. They are more about the design of constraints on societies than about factual description of an individual's human nature. Some economic stories look at first sight like scientific stories about the way in which wealth can be created and managed, and in their use of mathematical modelling; but they are really assertions of beliefs, usually unjustified except by Everyday or Authoritative Stories, or even merely self-consistent mathematical stories with little basis in everyday human nature. They have little to tell us about the roots of human nature, especially since the stories they tell about the fundamental nature of humans are often evidently incorrect.

Using Stories

Why do we need to distinguish these different kinds of stories, and what role will they play in an investigation of human nature? Each kind of Story provides a kind of explanation. (In future I will write "Story" with an upper case "S" when I refer one of these categories, and with a lower case "s" if I just mean any old story.) I want to use Stories (and stories) to give you a picture of human nature, situated in the natural universe in which we find ourselves. My choice of the candidates for most important kinds of stories is of course, biased. I will put the main emphasis on science and philosophy.

Sometimes stories seem to be related to one another like a ladder of explanation. Psychology, mental events in the lives of humans and animals, and the behaviour of individual humans and animals explains social behaviour. Physiology seems to explain psychology by talking about brain function; biochemistry explains physiology; and physics explains chemistry and biochemistry. Some people say that physics explains everything - indeed James Watson, one of the discoverers of the structure of DNA, famously said that only physics is real science, "and everything else is social work". We shall see that there are reasons to think that Scientific stories are much more interesting than that!

We tell ourselves stories to explain our nature. We are born. We live out our lives. We experience the world, make choices, and act in and on the world. Then we die and disappear. So what is it all about[9]? What kind of creatures are we and in what kind of universe do we exist? What *kinds* of Stories, and what *particular* Stories can best make sense of our lives? In the 8th Century St. Bede the Venerable compared human life to a bird that appeared out of the darkness in an Anglo-Saxon dining hall, was seen for a moment as it flew across the hall in the firelight, and disappeared again into the darkness. What kind of Stories could one tell about that bird?

---

[9] "I once 'ad that Bertrand Russell in my cab and I said to 'im, "Well, Professor Russell, what's it all about?" And do you know, 'e couldn't tell me!" Ascribed to a London cab driver.

My career as a scientist has left me very impressed by Scientific Stories. We shall see that they do not account for everything in human nature, and indeed there are some questions to which scientific stories contribute little or nothing at present, and perhaps never can even in principle. Puzzles that science leaves untouched may need philosophical stories; but there are some aspects of human nature that seem to be left nearly untouched even by philosophy. Despite all the advances in our knowledge there remain deep and profound mysteries. Many people feel a need for religious or other authoritative stories to speak about such mysteries: I do not. I am happy, if need be, to remain in doubt to my dying day.

This book is to help readers to examine their opinions, clarify what they imply for human nature, and decide what Stories to tell. All of us hold many opinions and beliefs, and tell ourselves many stories; but often we have never tried to see what we really mean by telling them. That is true of all people on any side of every controversy. I hope this book may help the reader better to explore his or her nature. As the ancient Greeks inscribed on the temple of Apollo at Delphi, "γνωθι σεαυτον"—"know yourself."

I will sometimes talk about the "biography" of a human. I mean something like the timeline of their life. By speaking of a biography we are not committed to say whether the event can be entirely described in physics, biology, psychology or some other language. So an event in the biography of a human could be someone saying something, reading something, meeting someone, thinking or making a choice, or the physiology of digestion, or electrical activity in the brain, or even the chemistry and physical processes at the level of cells or atoms in the body. *We want to understand what causes the events that make up a human biography.* But how many ways are there to ask, "*why*"? What would it mean to *explain* Salterella's jump?

PART 2

THINKING ABOUT THINKING

Chapter 2

Causes and Explanations

Felix qui potuit rerum cognoscere causas[10].

> Virgil. *Georgics*. No. 2.

Explaining metaphysics to the nation -
I wish he would explain his explanation.

> Lord Byron, *Don Juan*, 1819.

It would be nice to plunge straight away into what science has to say about life, consciousness, and other aspects of human nature. But these are difficult topics, and we need to be sure that we are using the right methods. It isn't as simple as one might think to look for causes and explanations. No wonder Byron was irritated by Coleridge.

Explaining Explanations

To explain events in a human biography we need to understand how different kinds of Stories are related. We want to know what was the *cause* of which the jump was the *effect*. If we can find a cause, we will have an explanation of the behaviour. The same applies to mental events, the action of the will, consciousness and the nature of life. What kind of Story best deals with the causes of human behaviour?

There are many ways to account for how and why a person does something. There may be several explanations for a single event. Three directions in which we might look are shown in Figure 2.1. First we have the physical sciences, the path running upward from the bottom of the figure, which describe events in Saltarella's body in the

---

[10] Happy is he who can understand the causes of things.

present. In our time it is common to regard Science as pre-eminent as a source of explanation. Secondly there are Saltarella's motives, interests, and life choices, the path running down from the top right, which in a sense are about the future as she sees it. Finally the social context and ecology of the person's life produce the path running down from the top left, and these are mainly Historical Stories.

People often think that one kind of Story gives a more profound, or more important, or more logical explanation than others. Some Stories are rational and others are only emotional or metaphorical. But then what do we mean by "rational"? Is it just another name for Science? Why should Science be so favoured? If people prefer some kind of stories to others for emotional or aesthetic reasons, not because they are true, does that matter? If I choose a particular kind of Story to explain an event, then I should also be clear whether other explanations are compatible with that story, or there is a contradiction between the kind of story I like to accept and other kinds of stories. Only if there is a contradiction shall I have to decide which Story to choose.

Let's start with a Scientific Story about the high-jump. Concentrate on the moment at which the muscles in Saltarella's leg drive her body upwards. It is certainly true that the cause of the jump is the sudden contraction of the extensor muscles of the leg, which causes the joints of the foot, ankle, and knee to swing open rapidly and with tremendous power, propelling her body off the ground. Muscles are made of many thousands of muscle fibres, each of which is a living cell which shortens rapidly and dramatically when the muscle receives stimulation from a nerve fibre. The contraction of the muscle as a whole, and hence the extension of the joint, can be explained by the twitches of the individual muscle fibres which make up the muscles and are attached to tendons and bones of the leg. The sudden shortening of the muscle fibres is caused by the biophysical and biochemical properties of the molecules myosin and actin, proteins which are found in the muscle cells. When an electrical impulse is propagated from the nerve fibres running from the brain to the muscle, the electrochemical properties of the myoglobin proteins cause one part of each of these molecules to slide across another, so that their overall length becomes less: the muscle fibre shortens[11]. When this occurs in thousands of muscle fibres within a fraction of a second the effect is a powerful contraction of the whole muscle. So there is a series of scientific stories going "upwards" from the molecular electrochemistry to the observable movement of the jumper that explains the jump. These stories give a causal explanation for the jump.

Now we could of course go down even deeper into the chemistry of living tissues[12]. We know that chemical properties are caused by the exchange or binding of electrons between atoms; and the properties of electrons and atoms and the constituents of the nuclei of atoms (the protons and neutrons), are described by physics, and in particular quantum physics. So can we say, as Watson seemed to say in the quotation we cited earlier, that the ultimate cause, the ultimate explanation of the jump is the quantum physics equations describing the dynamics of the nuclei and electrons of the atoms in the molecules of the muscle?

The answer has to be, "no", for several reasons. First, the same equations describe any atomic events whatever they are, not just muscle actions, not just a jump. The properties of any proton, neutron or electron are identical with all others of its species. Even if we were able to look at individual electrons we could not tell which ones are part of a muscle without also looking at the muscle itself as a whole. Furthermore, at the level of quantum physics there is no causality,

---

[11] P.Hoffmann. 2012. *Life's Ratchet*. New York. Basic Books.
[12] There was a time when the name *organic chemistry* was reserved for the chemistry of living organisms, because it was thought that there were certain chemicals that could only be synthesized by living systems. Following the successful artificial synthesis of urea, (a compound that occurs naturally in living systems,) *organic chemistry* was used more or less as a synonym for the chemistry of carbon compounds. And today, *biochemistry* refers to the chemical properties of living tissues, but with no implication that these are fundamentally different from any other kind of chemistry, save only that the chemical molecules involved are usually found in living plants and animals. In view of the enthusiasm for organic foods, materials, etc. it is ironic that the Greek word from which *organic* is derived originally meant *system*, *device*, or *machine*.

only probability. So how could we account for the fact that all the atoms in all the molecules of all the muscle fibres of all the muscles contracted in beautiful synchrony in exactly the temporal relation needed to produce the powerful extension of the leg which was part of the jump? Indeed, in other situations there are neural circuits that seem to be designed to prevent muscle fibres from

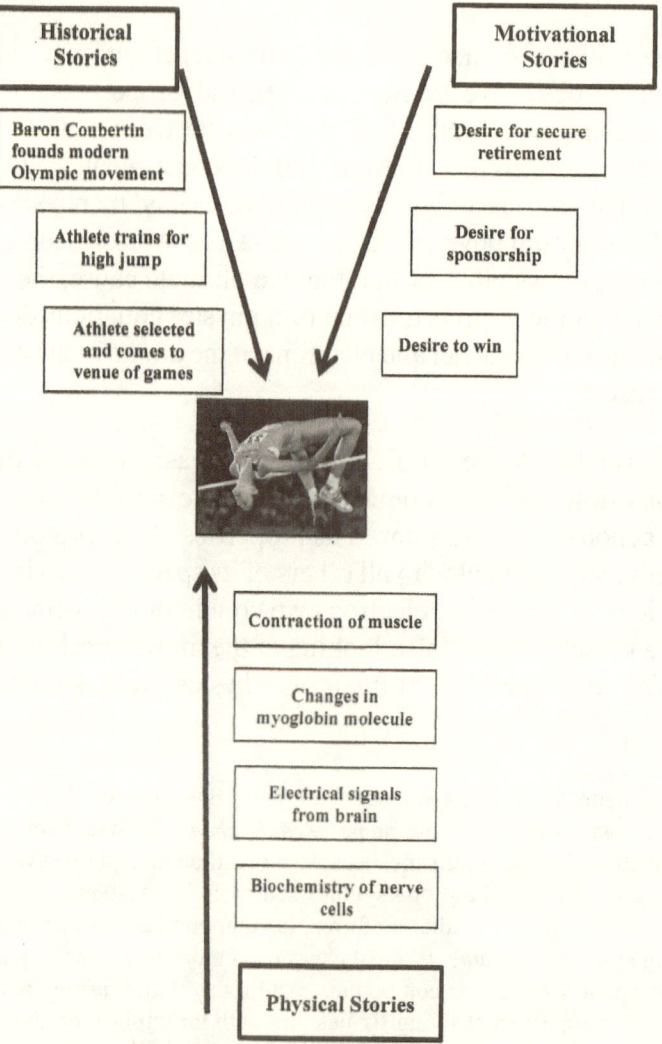

Figure 2.1 What caused Saltarella to jump?

contracting in synchrony[13]. Whatever we see when we look at the quantum events in the atoms of a myoglobin molecule in muscle we do *not* see a jump by an athlete; and given that quantum events are not causal but probabilistic, we *could not* see such events as *the* cause of the jump (although they might be *a* cause), because the probabilistic nature of quantum events could not guarantee the required synchrony. The probability of such synchrony among a huge number of independent quantal events is effectively zero.

On the other hand, it is certainly true that without these events the jump would not occur. And it *is* true that we can tell a story such as the following. The atomic events cause the molecular events that cause the proteins to change their position that cause a muscle fibre to contract that makes a twitch that with all the other twitches makes a muscle contract that causes a jump by the athlete Saltarella when she decides to do so. So in principle it seems that there is a complete causal account of the jump given by the scientific story. But how can that be, since we have just argued that it can't be the correct story? Why isn't the physical story complete? Well, Rose pointed out that in the case of his frog the story was incomplete without an "ecological" cause, namely the snake. The frog jumps *because* it has seen the snake; it jumps for the purpose of escaping from the snake. Ecological and motivational stories are needed for completeness. In the case of Saltarella the situation is even more complex. Just as there were several "downward" stories to explain the jump by relating it to the muscle twitch, so we can invoke stories "upwards" as in Figure 2.1.

What makes the muscle fibre contract is that Saltarella's brain sends electrical impulses down the spine through nerves to the muscles. What makes that happen at a particular moment? As Saltarella completes her run-up she sees and feels her position relative to the high-jump bar, and because of long practice her perceptions trigger the appropriate neural messages. So we should add a story based on

---

[13] The uncontrollable rhythmic shaking of a leg called *clonus* which a rock-climber experiences if he stands for too long supporting all his weight on a toe on a very small foothold is caused by the involuntary synchrony of muscle contractions. Renshaw cells in the spinal cord seem to have the function of preventing such synchrony.

the psychology of perception and cognition. But why is she there at all? She wants to win the high-jump at these Games; and the cause of her being at the Games is that she wants to win, and will make all the required effort, and suffer the pain of extreme physical exertion, in order to win. To win will gain her money and glory. The desire for victory causes the events that lead to the twitch and hence the jump. And we can even add a History Story: were it not for the fact that Baron Coubertin had founded the modern Olympic Games in 1894 Saltarella would not be able to compete; so there is a distant *historical,* or *ecological* cause of the jump.

How can several stories be true at once about a single event? What kind of a relation is there between Stories at different levels and of different kinds? There is no doubt that when anything happens in the brain atoms and subatomic particles change state. But we can't describe what is happening in the world seen at our everyday scale by describing the changes in electrons. Using the Atomic Force Microscope[14] we can watch the activities of complicated molecular systems inside a cell, and see what happens in, say, a ribosome or mitochondrion in a single cell. But the same events happen in many different kinds of cells when many different functions are performed. At what level of description do things differ in such a way that we can identify a perception, a thought or a decision?

Physicists often talk about events at the level of atomic or subatomic particles. But we can't talk about significant human events in such language. So should we talk at the level of molecules? At the level of the movements of ions in the neurons? At the levels of patterns of the firing of nerve cells? At the level of populations of neurons (what part of the brain fires)? Cognitive neuroscientists like to explain mental activity by the correspondence between neural activity in certain regions of the brain and mental events. Psychologists often explain mental events using computational models that simulate how people behave. Social scientists explain behaviour in terms of social forces and concepts involving culture, membership of social classes, and so on. Are all these different descriptions interchangeable? Can they be

---

[14] P.Hoffmann. 2012. *Life's Ratchet.* New York. Basic Books.

translated one into another? Or are some of them mutually exclusive? Are some "physical" and others "immaterial" or "spiritual"? Is there a real difference between the "natural" and the "supernatural", and does it help to explain what we do?

Different Stories provide us with different *kinds* of accounts of what happened in the biography of our athlete, and why it happened. They offer different kinds of causes and different kinds of explanations. Notice that all the accounts are *different*; all the accounts can be *true*; and *none necessarily excludes the others*.

That is almost always true. There may sometimes be reasons to say that a particular story is untrue, or inapplicable, and we may even want on occasion to exclude some kinds of stories as nonsense, or irrelevant, and argue that they should not be used at all to describe human nature. But the mere existence of several kinds of stories does not prevent any particular story from being true. Despite Watson's claim there may be reasons why it is more correct to describe an event in terms of social work than in terms of physics. Knowing about quantum physics will not help you to re-house single parents. But if that is true, why do we bother which Story people tell? Surely some are to be preferred to others? Is the acceptability of multiple stories perhaps due to the simplicity of the event? Well, let's try a more richly human story.

Suppose we want to understand why a woman gives money to a beggar. An Everyday Story might be just that in our society one should look after the poor. A Religious Story might be that the woman believes that God says that we should perform works of charity. An Historical Story might be that it is a tradition to give money to the poor on a particular day, because that was when King Wenceslas gave money to poor people. A Political Story might be that the rich support the poor in order to avoid revolution, and that the Government has recently cut the payments to the unemployed. A Scientific Story might be that the pattern recognition mechanisms in the woman's brain let her perceive the state of the beggar and also activated muscle movements which caused her to take money out of her purse and put it into the beggar's hand. A Philosophical Story might discuss whether, given the history of her education and the physiological determinism of the brain, it makes sense to say that the action was voluntary; did she really give the money at all, or was it conditioned behaviour? And of course fiction might say that someone was acting out a fairy story by Hans Anderson or the Brothers Grimm. So the availability of multiple stories is present here also, but their multiplicity is not worrying. As the philosopher Wittgenstein said, the meaning of a word is the use to which it is put: many uses, many meanings.

Looking back you may notice that we have not introduced several ideas that you may think are central to human nature. We have not spoken of minds, or souls, or anything spiritual. That seems strange. Should we perhaps have referred to Saltarella's intense "will" to succeed? What is going on? In Chapter 3 we shall see that just because you cannot measure something using physical measurements it does not mean that it does not exist – but *neither* does it mean that to be "non-physical" is to be supernatural or spiritual in the sense of non-material. An event can be non-physical in the sense of not being measurable by scientific instruments without ceasing to be part of the physical world.

Although usually several stories can be true simultaneously sometimes there may be real incompatibilities among them, and when that happens we have to choose which to adopt. For me the two most important kinds of stories are Scientific stories and Philosophical stories. The first is the most powerful way we have to add to our knowledge; the second is the most powerful way to clarify our thinking. Most thinking is done in words, and philosophy clarifies the way in which words carry meaning. We should only adopt beliefs that are supported by evidence and logic, and reject beliefs that are contradicted by evidence or are logically incoherent. And if something is uncertain, we must be prepared to live with uncertainty, and not arbitrarily choose one alternative just because uncertainty is disturbing. (We shall see that much knowledge is only *probable*, not certain; and that in many cases there can *never* be certainty even about the natural physical world.)

As an example of how empirical, or scientific investigation, could change a belief, consider an Everyday story. Many people believe that capital punishment will reduce murder, perhaps because they feel that it would certainly prevent *them* from committing murder. One way to see whether this belief is true is to subject it to scientific enquiry. If capital punishment prevents murder, than we would expect that in any country where it is abolished the murder rate would rise. That is a reasonable empirical prediction. In fact, there seems to be no country where the abolition of capital punishment was followed by a significant increase in the murder rate. So the original belief is untrue and we should give it up, however surprised we are by the result of the investigation. On the other hand, science cannot tell you directly whether you have free will, because it is not clear what predictions we could make that could be tested using scientific methods. Before using science to examine free will we have to define clearly what we mean by *free* and *will* and whether these can be recognised as physical events. The activity of clarifying definitions is a matter for philosophy, not simply for science.

When we discover causes we also arrive at explanations, and science discovers causes. So we need to understand something about the nature of science, what kind of certainty we can have about our knowledge of the world, and what it means to have probable rather than certain knowledge. But first we need to clarify what we mean when we say one thing is the *cause* of another.

What is a cause?

Over 2000 years ago Aristotle noted that there are several meanings of *cause,* and this was well known in the Middle Ages. Today there is a tendency to think that there is only one kind of cause, sometimes called a "billiard ball" cause. If I hit a ball with my billiard cue it moves, and the physical impact of the cue is the cause of the movement of the ball. That kind of cause covers many events in mechanics, chemistry, physics and even biology. But it does not cover all causes. Perhaps it's time for a fable.

How to Drive a Car: a fable.

Someone is driving a car and decides that he needs to go faster in order to catch a train. Let's ask the question, "Why did the car go faster?" We can give several causal accounts, each of which is true.

1. The speed of the car increased *because* the driver wanted to catch the train.
2. The speed of the car increased *because* the driver pressed down on the accelerator.
3. The speed of the car increased *because* the more fuel was burnt in the cylinders so that the force on the pistons increased.
4. The speed of the car increased *because* the reading on the speedometer rose.

It is quite natural to use the word *because* in each of these sentences. Both (2) and (3) are "billiard ball" kinds of causality that today are the most common Everyday Story about causes. These are the kinds of cause that the philosopher Hume discussed in the 18th century when he claimed that to identify a cause was simply a matter of noticing that two things always occurred together. Whenever water is sufficiently heated, it boils. Whenever a cue hits a billiard ball, the latter moves. In our case, when the number of molecules of petrol that are oxidised (burnt) increases so does heat released in the combustion chamber and so the pressure on the pistons increases. So, heat is the cause of water boiling; the movement of the accelerator is the cause of more hydrocarbon molecules being burnt; and the energy of the oxidation of hydrocarbon molecules is the cause of pressure on the pistons.

Notice that even if Hume was right and when something causes something else they always occur together, the reverse is not necessarily true. Correlation does not imply causality. If red-haired people are short-tempered it is not the colour of their hair that causes their bad temper, nor the opposite. This is one of the problems about epidemiological research. If people who eat eggs have high cholesterol that does not guarantee that eating eggs is the cause. It may be that the kind of people who eat many eggs have some other characteristic of their life style that gives rise to the increased cholesterol, or that there are many different and unrelated causes that contribute, of which eggs are trivial, although the two go together. This is why the tobacco industry could argue for many years that smoking did not cause cancer, although in that case the correlation *did* in fact reveal a causal relationship that was confirmed experimentally. If electrical activity occurs in some part of the brain whenever we dream, which is the cause and which is the effect? Or is that not the right question?

We should take seriously the fact that the four sentences in this fable all seem both sensible and natural, and must reflect some real aspect of what we mean by *cause*. Aristotle gave them names, which we translate as *final* cause, *efficient* cause, *material* cause, and *formal* cause. Let's look at his analysis more closely, because we will find later that it is very useful to make these distinctions when we are trying to understand human biographies.

Final cause.

Final here doesn't mean the last cause, but the *end* in the sense of *purpose* or *goal*. Sometimes this kind of cause is called *teleological* or in modern writings *teleonomic*, from the Greek word τελος, telos, which means *goal, end,* or *purpose*. In most Scientific Stories it is said that we should not talk about purpose. But is is interesting that Norbert Wiener, who founded the science of cybernetics, argued that cybernetics is really the science of purpose and final causes in animals and machines[15]. In the car the final cause of the increase in speed is the driver's goal of catching the train. Rose's frog jumped in order to escape the snake. Saltarella jumped in order to win the high-jump.

Efficient cause

An efficient cause is one guaranteed by experience or rules, without any deep understanding of why it works. "Don't worry about why it works. Just do this and all will be well." In this case the driver

---

[15] N.Wiener. 1958. *Cybernetics: communication and control in the animal and the machine.* Cambridge, MA. MIT Press.

has learned to control his speed by pressing the accelerator, even if he does not understand the underlying physics or mechanics. He works with efficient causality. In the USA many experienced nuclear power plant operators objected to having to take college courses on nuclear physics, claiming that due to their extensive experience they were experts[16]. In effect they were saying that their training had given them a sufficient understanding of the plant based on efficient causality. Most people understand how their their cars work in terms of efficient causality. Rose's frog escaped because it jumped. Saltarella jumped because her training made her identify her position in her run-up as being optimal for her take-off.

Material cause

A material cause involves an appeal to deeper, usually physical, understanding. An action is effective because by performing it a physical or chemical process that underlies the apparent events is brought into play. In driving a car, the cause of the increased power is the increase in the rate at which matter is changed into energy. Material causes are often part of the underlying Scientific Story of why something happens. Although they did not say so in so many words, it was their belief that an understanding of material causes would help operators to handle abnormal plant states during accidents that led regulators to demand that all operators of nuclear power plants should have university degrees. Burning petrol was the material cause of the speed of the car. The muscles and chemicals in the legs of Roses's frog were the material cause of its jump. The biophysics of the muscle cells in Saltarella's legs were the material cause of her jump.

---

[16] This observation is based on conversations during my interviews of such operators when I was investigating the human factors of nuclear power plant safety during the late 1980s.

Formal cause

The notion of formal cause is very important for Aristotle's account of causality and particularly for his ideas about life and souls. But it sounds odd to modern ears. Why should we say that the increase of the speedometer reading *causes* an increase in speed? The reason is, as the adjective implies, *purely formal*. That is, it is not about physics or chemistry, and it is not about practically effective tactics that the driver has learnt. It is about how we use language to describe what has happened. So we say that because of what we mean by the words *energy* and *speed* and *speedometer*, the occurrence of a higher reading on a speedometer makes us classify the state of the system we are measuring as going faster. The higher speed reading causes us to describe the system as faster, because otherwise what we say about it will not be consistent with other ways we use the words involved. The fact that Rose's frog left the ground due to a rapid extension of its legs means that we can describe it correctly as having jumped. Saltarella jumped because she was a competitor.

Although we don't often make them explicit, Aristotle's distinctions are very useful. A modern example of distinctions between kinds of causes is that of Pinker[17], who might call accounts of Salterella's jump using references to atoms and molecules *ultimate* causes, and accounts talking about her past history or her approach to the bar in her run-up *proximate* causes.

Let's go back to Saltarella. The *final* cause of the jump is her desire to win a gold medal. The *efficient* cause of the jump is her recognition that she in exactly the right position for her take-off (due to her past training of the required movements). The *material* cause is the set of events in her nervous and muscular systems. The *formal* cause is her being a contestant in a high-jump competition. You may like to map these distinctions onto parts of Figure 2.1.

It may seem that it only makes sense to talk of *final causes* in the biography of creatures that have something like human language.

---

[17] S.Pinker. 2002. *The Blank Slate*. Penguin Books.

Often what looks like a final cause, for example a cat sitting by a mouse-hole, can be re-described as an efficient or material cause, although sometimes we can correctly claim that the lives of animals without language can show final causality; remember that Rose's frog jumped to escape from the snake. On the other hand there is a warning in the behavior of certain insects that seem to seek out dark places to hide. Such behaviour was first described as being *negatively phototropic*, a purposive avoidance of light, a *final* cause. Closer research showed that when light falls on them they move in random directions. When no light falls on them they stop moving. So what looks like purposive behaviour, namely the seeking out of dark places, is just the result of random movement in light and absence of active movement in the dark. Only material cause, a response to light, is involved: there is no final cause. But it is important to recall that when Wiener developed the mathematics underlying cybernetics, and in particular those dealing with what came to be called "negative feedback", he argued very strongly that cybernetics should be thought of as the science of teleology, of purpose, both in the animal and the machine. Very early in the era of cybernetics the neurologist Gray Walter built a goal-orientated electro-mechanical "tortoise" which sought out other such "animals", approached lights, and withdrew when touched: and today we see the mature technology of such purposive behaviour in robots, although we do not imply that the robots are conscious of their goals. *Material cause*s usually refer to "deeper" or "underlying" mechanisms, such as the activity of muscle fibres or biochemistry that cause the jump. Scientific explanations are typically given in terms of material causes. *Efficient causes* are usually invoked at the interface between the person who performs the action and the properties of the environment in which she acts. (Scientific explanations are sometimes of this kind.)

So do the classical questions about mind, will and soul refer to different kinds of causes than those used by science? Are there important differences which should make us look for different *kinds* of explanations when we discuss aspects of human nature? What are the special roles, needs and characteristics of philosophical and scientific Stories in causally explaining human nature? Indeed, what is it like to do philosophy and science?

Chapter 3

Ways of Thinking: Philosophy

> Philosophy is a battle against the bewitchment of our intelligence by means of language.
>
> Ludwig Wittgenstein[18]

> Philosophy is. . . . an activity, the activity of analysis, the activity of making thoughts clear and sharply bounded. . . . More precisely, the activity of analysis applied to non-philosophical propositions – to the propositions of everyday speech. . .(and). . . to the propositions of natural science – makes them sharp.
>
> Anthony Kenny[19]

Fundamental Words

Probably the most widely held Everyday Story about humans goes something like this. Humans are living creatures, and what distinguishes living from non-living creatures is having a *soul*: but not many people ask themselves carefully what they mean when they use the word. Some people say that having a soul is what distinguishes humans from all other living creatures; others that it is the kind of soul that humans have that distinguishes them from other creatures. Many people believe that humans continue to exist after they die, and that the soul continues to exist after death because the soul is "spiritual" whereas the body is material. They say that while people are alive they are made up of body and mind, or body and soul. The body, in particular the brain, is physical or material, while the mind is

---

[18] L.Wittgenstein. 1968. *Philosophical Investigations.* Oxford. Blackwells.
[19] A.Kenny. *Wittgenstein.* 1973. Harmondsworth: Penguin Books. P.101

"spiritual" rather than physical. The soul is the "real me". So human life has two aspects, physical life comprised of events in the material body, and mental life comprised of events that occur in the mind. Typical mental events include consciousness, perception, thinking, and acts of will. Many religious people say that each human soul is created by God, the life of a person is sustained by God, and it is to God in His heaven that the soul goes after death. Others say that at death the soul migrates to a new body and then life begins for that new individual.

Talking like this in terms of souls and minds is certainly the most common Everyday Story of our time.[20] But does such a Story make sense these days? Where does it fit into Saltarella's jump? We know more and more about the nature of living systems and about the function of the brain. We have new and powerful ways (such as fMRI[21] machines and the Atomic Force Microscope) to watch the action of the brain and to analyse events in the components (nerve cells, neurons) of which the brain is composed. We understand more about biochemistry and chemistry than ever in the past; and underlying chemistry is our understanding of physics, of the properties of molecules, atoms, subatomic particles and radiation that are the fundamental components of the natural world. Increasingly many people find it likely that Scientific Stories, rather than Religious Stories, give the most satisfactory account of causes in a biography of human life. But even when we are told something that seems straightforward (that a particular part of the brain is used in decision making, or that intelligence is affected by our genes, for example,) we should begin by asking what those discoveries mean. What is a decision? What is intelligence? How do we examine the effects of genes? As we start thinking, we have to think about thinking itself. That is the purpose of *philosophy*.

Let's list some words that we seem to *need* to describe human nature, words that almost force themselves on us. Let's call them

---

[20] If we add up Christians, Muslims, Buddhists, Hindus, Jews, etc., it is clear that most people in the world still believe Religious Stories.

[21] functional Magnetic Resonance Imagery

Fundamental Words, because they seem to apply to human nature in a particularly obvious way. They seem almost impossible to avoid if we want to capture the special qualities that we associate with being human. It is not so much a matter of proving they exist, more that each of us needs them to talk about his or her own daily life. Fundamental Words have an aura of mystery; but at the same time they refer to aspects of our nature that are immediately obvious, and some of which make us fundamentally different from other entities that inhabit the universe, whether alive or not. The words are

Life
Intelligence
Consciousness
Mind
Will
Self
Soul

Why emphasise these words? There may be others that seem equally important in defining human nature, words such as *love*, or even *politics*. But I have found the *Fundamental Words* are ones that occur most commonly in discussions about human nature, particularly when people worry about what to make of scientific discoveries. If we are not alive and conscious in the first place, then other questions, such as those about love or politics, cannot arise. Fundamental Words are not special in the sense of having secret powers like those in fantasy novels, such as Tolkein's *The Lord of the Rings*, the novels of Charles Williams, or Harry Potter books; but they play a very special role in Everyday Stories about human nature. There is something that they capture about human nature that has made people use words like them throughout history. They name qualities or characteristics that we all know that we possess. They sound as though they refer to special parts or components of humans. That perhaps is their power. We have seen that to find a cause for something is often, perhaps usually, what we mean by an explanation. So to explain the meaning of the Fundamental Words, to show what kind of causal story applies to each, might lead to an explanation of human nature, in the sense that it would leave us feeling, "Ah! So *that* is what it means to be human. *That* is why Saltarella jumped."

If the Fundamental Words raise difficult questions, what kinds of stories do we expect by way of answers?

- What is the difference between living and non-living things? *What causes something to be alive?*
- What does it mean to say a person has a soul? Where does the soul come from? *If you think that people and perhaps animals have souls, how does a soul cause a person to be who and what he or she is?*
- What is the nature of mind? How do mind and body interact? *How does the mind cause the body to move and how does the body cause the mind to have thoughts and know about the world?*
- Are members of different races more or less intelligent than one another? *What is intelligence? What is race?*
- Do we have free will? *If I have a brain that is a biological mechanism, how can I avoid being made (causally) to do things by the state of that mechanism? How does my will make my body act?*
- What is the nature of consciousness? *How does the matter of the brain make me conscious of the world?*

Even if we think that science has made some of these questions obsolete, it is difficult to discuss human nature without considering at least why that is so. If you say, for example, that you don't believe that humans have "souls", while I say I believe that they do, we had better make clear what each of us thinks the word "soul" means, or our discussion will be completely vacuous. (In fact we might even agree if we mean different things by the word!)

I think that a materialist account of the universe, including humans, is correct in one sense, but wrong in another. I don't think there are spirits, ghosts, or angels with which we share the universe. I don't think that we continue to exist after we die. But it is not easy to prove all this, and at the same time I think it makes sense to talk about life, minds, free will, consciousness, and the dignity of human beings. A materialist account of nature does not destroy human values, although the nature and origin of such values is to some extent unclear: perhaps we create values as we live. Science is not happy to talk about values. There are aspects of human nature that transcend a materialist account, but do not require us to think there is a non-physical part of human nature. There are aspects of human nature that cannot be described by the physics of a person such as Saltarella, but do not imply a "Ghost in the Machine"[22].

One way to understand Fundamental Words is to understand relations among Stories, and in particular between Scientific Stories and Philosophical Stories. We will see later what it means to tell a scientific Story: we can only use science to examine a well-defined phenomenon, and moreover one that's amenable to scientific analysis. (Not all things are.) And to decide whether something has a clear definition, to decide whether the way we think about it is clear and consistent, we need a different discipline, philosophy. It is philosophy, not science itself, that clarifies how we think, examines the implications of language for definition, and provides a description of what science itself is. Philosophy helps us to ensure accurate descriptions of phenomena, definitions of phenomena that allow science to be used to examine them. Modern tools have opened up very powerful new ways to do research, but consider the following remarks by Roger Scruton about the study of brains:

> There are many reasons for believing the brain is the seat of consciousness. . . . (Neuroscience) cannot fail to encourage the superstition which says that I am not a whole human being with mental and physical powers, but merely a brain in a box. . . (The invocation of neuroscience) perfectly illustrates the prevailing academic disorder, which is a loss of questions. . . .Human beings are a biological concept, whereas "persons", who do most of the things in which we are interested, are not. . . .We understand people by facing them, by arguing with them, ... We do not understand brains by facing them, for they have no face. . . . We should recognise that not all coherent questions about human

---

[22] A phrase invented by Gilbert Ryle in his book *The Concept of Mind* to capture the way in which dualists think about the relation of mind to body.

nature and conduct are scientific questions, concerning the laws governing cause and effect.[23]

Now, is this profoundly true, or profoundly nonsensical? What kind of account should we give about how science and philosophy relate to each other? If we want to explain the causes of human nature, perhaps we had better look again at what we mean by *cause* and *explanation*. When we begin to look carefully at how Stories relate to each other, it is not even clear that ways of talking we usually think of as being completely in conflict with each other are necessarily so.

The Electrician and the Advertisement: a fable.

Once upon a time a man owned a restaurant. One day a visitor came by and sold him "an advertisement" that he could put outside his restaurant, and which the visitor asserted would bring him many more customers because it would tell people that here was a restaurant. It would light up at night and people would be attracted. The owner waited until the evening and then plugged it in and switched it on. No more people came.

After a few days he took it to an electrician. "Please look this over and tell me why it is not working.", he said.

The electrician took the device, and examined it carefully, using all his scientific instruments. He checked out the wiring, and made sure it was correct. He measured the electric current at all the junctions and through all the wires, measured the amount of light leaving the tube that contained gas, and even managed to establish that the gas in the tube was neon, by using a spectroscope. Everything seemed to be working correctly, the current was flowing correctly, and the voltages were correct throughout.

The electrician sent for the owner and told him that he had thoroughly checked out the apparatus, and everything was as it should be. "Look," he said, "I will show you how the electricity passes from

---

[23] R.Scruton. *The Spectator*, 17 March 2012. *Brain Drain*

one component to another, and how it makes light appear in the tube at the end." He did so. "There you are," he said, "it works perfectly. Now you know how it works, and how it produces the light."

"But," said the owner, "the man who sold it to me said that it is an advertisement. Where is the advertisement?"

"What do you mean?" said the electrician. "I'm an electrician, and there is no such thing as an "advertisement" in electrical engineering. I have shown you everything that happens in the system. There is nothing else that I can do, although I am a very good electrician. I don't know how to measure an "advertisement". What is an "advertisement"? None of my measurements detected an "advertisement'. In fact, I'm pretty sure that according to my instruments there is no such thing to be measured. The instruments certainly did not indicate that there is anything other than the electrical and physical components."

You see, they lived in a strange culture. Although their knowledge of electricity was quite advanced, and the claims that the electrician made as to his ability were correct, the culture had no concept of "advertisements", and such things were unknown and not used in the culture. The problem was that although they knew *everything* about how the electrical system worked, and had a complete description of its properties, no electrician could measure the efficiency or even the presence of an "advertisement".

The physical description of the system was *complete*, but of course an advertisement is not completely described even by a complete physical description of the structure and function of the sign. It is not that an advertisement is supernatural, or transcendent, or spiritual, or ghostly. But it is non-physical in the following sense: for something to be an advertisement people must live in a culture that understands the idea of an advertisement. To be an advertisement a thing must invoke a cultural response, not just operate as a physical system made of matter and energy. The neon light would have to be in the form of the word "Restaurant", and even then would only function for those who knew the language and realized which word was displayed. I once experienced a version of this fable. I was working in China, and after several weeks of teaching at the Huajong College of Science and

Technology I went for a sightseeing trip by myself, without being able to speak any Chinese. On my first evening in Suchow I went out to look for a restaurant and suddenly realised that I had no idea what is the appropriate Chinese character. I stood in the street looking at the shop fronts, completely unable to respond to what I assume were functioning advertisements. I could not recognise advertisements (if that is what they were!). So the parable is not so far-fetched.

The point of this fable is that a complete account of the physical working of a system does not prevent it having additional properties that are not covered by the physical scientific description. There may be several descriptions, each complete, which together cover *complementary* sets of properties. To measure whether a neon sign advertisement is working, it is not enough to measure voltages and currents and light output. Rather, you must measure the way in which people change their direction of walking and the probability that they will enter the restaurant with and without the sign. Neither description precludes the other, although the purely electrical description is sufficient to see whether the neon sign works as a source of light. The advertising description will only be valid if the physical description is valid, but to tell which account is appropriate you have to ask questions in an appropriate way. *A scientific description is not always a complete description even of a physical system.*

Another example of complementary descriptions is of course the language of lovers. A lover who is a physiologist or anatomist can give a complete biological description of an eye but may nonetheless describe his lover's eyes as being "stars in a pair of clear, blue pools" - and be correct!

Doing Philosophy

How should we approach ideas such as the *Fundamental Words*? We may want just to sit and think about them, or we may want to use science to examine them. In either case we must try to define what we are talking about. Most thinking about abstract ideas requires words, and philosophy clarifies the way in which words carry meaning. For an imaginative picture of what it might be like to think without our

Science, Cells And Souls

kind of language, read *The Inheritors*, by William Golding[24]. For a more recent scientific picture see the treatment of "fast thinking" in the book by Daniel Kahneman.[25]

Often a lack of clarity in how we use words can make a hard problem even harder. People may say that philosophy is "only" about words, or "just playing verbal games", while science examines real characteristics of the physical world; but it is important to understand that if a Scientific Story is told in the wrong words, it can be confusing or misleading even if the empirical facts are certain. Remember the Electrician and the Advertisement.

To think clearly about human nature needs philosophical analysis. Consider three problems.

1. If we want to know whether it is possible to create life, we need to be clear about what it is that we would be trying to do. *What do we mean by life? What criteria do we use to decide whether something is alive or dead?* If we don't decide on criteria, then no experiment can answer the question, because we won't know whether or not we have succeeded.
2. *Is there such a strong genetic component in intelligence that there are real racial differences in intelligence between races?* Well, what do we mean by *race* in this context? And what do we mean by *intelligence*? Only if we can first answer those questions can we go on to do the relevant scientific investigations.
3. If we watch someone thinking about problems while she is in an fMRI apparatus parts of the brain "light up" when she thinks. Does this mean that we are seeing *thinking*, and that is the part of the brain with which she thinks? *What do we mean when we say that someone thinks? Can thought be measured by electrical activity? Or do we mean by a mental activity something that is not physical and hence can't be measured electrically?* But if so, what is it that we see in the

---

[24] W.Golding. 1955. *The Inheritors*. London. Faber and Faber
[25] D. Kahneman. 2011. *Thinking Fast and Slow.* London. Penguin Books.

fMRI investigation? And what does it mean to claim that there is an event that is "not physical"? Is thinking like an advertisement?

One point is very important. Philosophy can help us to make our problems more precise, more well-formed. But that is not the same as demanding an exact definition. Consider the word *life*. We are not trying to decide what life *really* is, what is its precise definition. The philosopher Plato seems to have thought that for any idea there was, somewhere, an exemplar of its essence, what it really is, and that that is what we mean by the definition[26]. Here we are not thinking like that. From one moment to another, from one way of discussing a problem to another we may find that we want to use a slightly different definition. As we will see, science in particular often uses what we call *operational* definitions to avoid the Platonic problem. But for each way we think about an idea we must make its use, for that purpose, clear; and make clear how it relates to other occasions when it is used. In a later chapter we will look at the relation between names and what they name. Here we want just to see how in general we can benefit from philosophy. As the philosopher Wittgenstein said, the various meanings of a word are like family resemblances. The members of a family are not identical, but are clearly all members of the same group.

Here is a quotation from a book on the philosophy of mind to show you how a professional philosopher thinks.

> "The materialist is right in claiming that to describe a state of mind is to describe, at a certain degree of abstraction, a physical object. But the physical object which is described by mentalistic predicates is a human being, not a human brain. The brain states characteristic of speakers of English – if we assume, with the materialist, that there are such – may, for all we know, be reproducible *in vitro*. However successfully they were reproduced they would

---

[26] The novel *The Place of the Lion* by Charles Williams describes what it might be like if the Platonic Forms actually appeared in the world one day.

not constitute knowledge of English; for it is people, not brains, to whom it makes sense to attribute such knowledge."[27]

The writer is drawing our attention to what is implied by the way we use language to think about the mind-body problem. He is saying in effect, "If you look at the way in which we actually talk about minds and bodies – as we have been doing in this book – you will see that what I have just said must be true of how we speak. If we start saying that it is the brain, rather than the person, that knows English, then we will find that there are all sorts of self-contradictions and muddles that will arise in our account of the human mind. In order that everything that we have said about minds shall make sense we have to say that a person, not a brain, is what knows the English language." And he might then go on to say that looking into the electrical activity of the brain is not the way to tell whether someone knows English. After all, the best way to know whether someone understands English is to have a conversation with him in that language. Knowing facts about the electrical activity of the brain won't make it any more certain. So whatever fMRI studies investigate, logically it can't be simply *the nature of knowing English* (although it might be what happens in the brain when a person knows English).

Many ideas in this book need philosophical analysis to clarify them, particularly the *Fundamental Words*. Scientists tend to be contemptuous of philosophy, and think of it just as playing verbal games; but that isn't true when philosophy is used correctly. It can both help to design good scientific research and make clear what the findings of that research mean.

There are many things said about humans in Everyday Stories and in some Religious Stories that do not seem to fit with Scientific Stories. If they *really* clash with Science then I would want to say that they must be abandoned, because as we shall see science is very strongly supported as a way of understanding human nature. But it may be that if we examine carefully both what science says and what the

---

[27] A. Kenny. 1989. *The Metaphysics of Mind*. Oxford University Press. P.151.

other stories say we will find that in some sense both can be true. Remember our first story about what caused Saltarella to jump. Are there perhaps *many* stories that are true? Philosophical analysis can help us to decide whether it would make sense to say so. Some people say that what makes a person alive is having a human soul. Others say that it is "nothing but" a matter of chemistry. Are these really alternatives, or are they in some way saying the same thing?

In Everyday Stories and in Religious Stories we often hear that humans are "not just" matter. There is something about them that is of a different nature, sometimes said to be "spiritual", or "immaterial", or "transcendent". And in some Stories it is that part of a person that is "the real me" and even survives death. Well, what do those words mean? Let's look more closely.

What does it mean to say that something is *material* or that it is *spiritual*, to adopt a philosophical standpoint of *materialism* or *spiritualism*? By *spiritualism* I don't mean the existence of ghosts, or what happens when one tries to summon up spirits at a séance, but the claim that one can't give a complete account of the natural world purely in terms of matter and its properties; that we also need to speak of some other kind of properties that are "nonphysical", or "transcendental".

In the book quoted above Kenny implies that if something is material it has length, breadth and height; it can be *located*. That is not quite enough. For a start modern physics demands that if something is material, it is defined by *four* properties, length, breadth, height and time. Let's call these x, y, z, and t. Then for many purposes we can say that a material object is one that has a name, for example "stone", and has dimensions $\{x,y,z,t\}$ that locate it in a particular place *and* time[28]. Another difficulty is suggested by Ashby[29]. He

---

[28] To be strictly accurate, located in *space-time* as defined by Minkowski. A careful reading of Kenny shows that he makes the same point, but at a different place in his text. String Theory claims that we need almost a dozen dimensions to describe the physical world.

[29] W. Ross Ashby, 1956. *Introduction to Cybernetics.* London: Chapman and Hall.

shows that cybernetics can be applied even to immaterial systems providing only that we can identify accurately the states they can be in. He shows how to bring peace and quiet to a haunted house by going through an appropriate set of operations in the right order, even though ghosts are, by definition, immaterial. So if there were such things as ghosts[30], let's say a traditional "white lady", that ghost as seen by an observer would have length, breadth and height and would appear at a particular time. Therefore it has a four dimensional description, even though if you try to touch it your hand passes through it. So if ghosts exist and are examples of nonphysical entities, then the simple description in terms of $\{x,y,z,t\}$ is inadequate to define materiality. The difference between *material* and *immaterial* is not as straightforward as it seems.

Let's try another approach. Why is anyone ever tempted at all to think that nonmaterial entities, perhaps *minds*, exist and are part of a human when the universe is so obviously material? I know that I have a mind because I know I can do mental arithmetic; but I don't stop to consider whether it is a physical or non-physical thing. I just do things with it. Or rather, more correctly, I just do mental things.

There certainly are things we encounter in human nature that are not material. For example, numbers are not material. (I don't mean written numerals on the page, where the marks are clearly physical, but the numbers that are the concepts of mathematics). Kenny gives the example of "ability" as being a word that does not refer to a physical object.

> An ability.... has neither length nor breadth nor location. This does not mean that an ability is something ghostly: my front-door key's ability to open my front door is not a concrete object, but it is not a spirit either.[31]

It is quite easy to think of a whole range of things that are not physical. But this does not mean that they are made of some nonphysical

---

[30] Let me be clear that I do not believe that ghosts exist.
[31] A. Kenny. *The Metaphysics of Mind*. 1989. Oxford: Oxford University Press. P. 72

material, merely that there are things to which the distinction between material and immaterial, or the language of physics, are inappropriate.

These puzzles about the relation of material to non-material things have well-known historical origins. One is the pervasive influence of *dualism*, particularly in the writings of the seventeenth century philosopher Descartes, which has strongly affected our Everyday Stories even though most people don't know where the ideas originated. The assumption of most Everyday Stories is that the *real me* is a nonphysical entity somehow contained inside me; what the philosopher Gilbert Ryle delightfully called "the Ghost in the Machine", and that I shall shall henceforth refer to as a GIM for simplicity. If one can believe in such an entity, it offers some hope that when the body ceases to exist that entity will continue to exist. Hence it opens the way to belief in survival after death, which many people find comforting. There is also the folk tradition of the existence of ghosts, spirits, and the kind of things that appear at Spiritualist séances. I can't take these very seriously, and anybody who is impressed by séances should read *The Road To En-dor*[32].

But why in the first place is there a temptation to think that there is a part of a human that is not material, such as the soul or the mind? Why do we, in Ryle's words, think there is a "ghost in the machine", a GIM? And, perhaps more importantly for our purposes, what would we lose if we gave up the idea? What properties of humans would we not be able to understand? In what follows I will use the word *ghost* or GIM as Ryle used it, as a shorthand way of talking about any hypothetical part of me that is not physical but immaterial. (That will avoid the need, whenever I mention this "part", to say, "soul, or mind, or whatever".)

---

[32] E. H. Jones. *The Road to En-dor.* 1942. London: Bodley Head. The true story of a pair of WW I British prisoners of war in Turkey who not merely convinced their fellow-prisoners of the reality of spiritualism, but also persuaded their gaolers to hunt for buried treasure in Mesopotamia, and had themselves repatriated on the orders of the "spirits". When they told their co-prisoners after the war how they did it, some of the latter would not believe them and insisted that the spirits must have been real.

Believing in a GIM would not actually help us very much. Firstly, it would not help us to understand why we are conscious, as distinct from just behaving. If the real me is not my body but my GIM, as Descartes tried to show, if it is my ghost that has consciousness, and if my ghost is a part of me that can really exist even if the body stops functioning, then there are many puzzles. Why does consciousness disappear when I am hit on the head with a brick? Why, to be less flippant, do I cease to be aware of my existence when I am deeply anaesthetised? Surely in the latter case *I* would still be aware of existing, because that is almost a paradigm case of what Descartes had in mind – *I* have lost a normal functioning brain but *I* am still there. However, that is exactly what does *not* happen. Under deep anaesthesia I cease to be aware of my own existence. In that state I do not know whether I am alive, or whether I exist. The medical team in the operating theatre are better placed to know whether I am alive, whether *I* exist, than *I* am!

Secondly, to believe in the GIM does not throw any light on how *I* can do things. What kind of cause would make a non-physical part of me interact with my physical body? If my decisions and my acts of will take place in a non-physical part of me, how can they result in movement, make my muscles twitch? Think about Saltarella. In Chapter 2 we saw a series of stories about the causes of her jump that went backwards and forwards in time, and upwards and downwards in physical mechanisms. But we did not need to talk about souls or wills or minds. Is it strange that there was no need at any point to involve the "ghost" as a cause? Given the enormous importance of the will to win, and the conscious decision to make the supreme effort to clear the bar, why did the story not have to mention of the "ghost" of the jumper?

If there is no "ghost", immaterial "soul", "self" or "spirit" in my body that is the "real me", then *who am "I"*? I think that we are here being misled by language. As Kenny[33] says:

---

[33] A. Kenny, *Metaphysics of Mind*, 1989. Oxford University Press. Page 87.

> At one level, 'the self' is a piece of philosopher's nonsense consisting in a misunderstanding of the reflexive pronoun. To ask what kind of substance is my *self* is like asking what the characteristic of *ownness* is which my own property has in addition to being mine. When, outside philosophy, I talk about myself, I am simply talking about the human being, Anthony Kenny; and my self is nothing other than myself. It is a philosophical muddle to allow the space which differentiates 'my self' from 'myself' to generate the illusion of a mysterious metaphysical entity distinct from, but obscurely linked to, the human being who is talking to you. The grammatical error which is the essence of the theory of the self is in a manner obvious when it is pointed out. But it is an error which is by no means easy to correct, that is to say, it is by no means easy to give an accurate account of the logic, or deep grammar, of the words 'I' and 'myself'.

That passage expresses the essence of the problem in the language of a professional philosopher. The main point is that sometimes what is misleading us about the nature of reality is not reality itself but the structure of language, which may makes us think that 'my *self*' is different from '*myself*'. And if we are misled about what we are made of, then we will necessarily be misled as to what our scientific experiments are examining. We will certainly find it difficult to give a causal account of being human.

It may be easier to see the point informally, so here is an example called,

The Philosopher's Breakfast: a fable.
"Good morning, John. What did you have for breakfast?"
"I had a fried egg and a sausage."
"That sounds good. Who ate it?"
"I did."
"Was it really *you* who ate the fried egg?"
"Of course it was! I just said so!"
"But when we were talking philosophy yesterday you told me that "the real you" was not physical, but the immaterial part inside you.

So was it your spirit who ate the egg? That's rather surprising: I didn't realise that spirits could consume fried eggs, and yet the egg has clearly gone."

"Don't be silly. *I* ate it!"

"But I thought you said *you* are not really your body, but your "self" or your "soul". Perhaps only your body ate the egg, and *you* did not really eat it."

"You're being ridiculous! Of course I ate the egg! It was delicious!"

"Well then, *who* ate it? You say it was not eaten only by your body; and that it was not eaten by your spirit."

"I tell you *I* ate it! *I*, John, the person who is talking to you."

What hungers, what eats, what talks is a *person*, a human being, not a Cartesian ghost or *Ego*. What lives is a person. What dies is a person. And when a person dies we are left with an inanimate body not because something has left it, but just because it has stopped being a person. That is why surgery is so dangerous: in deep anaesthesia we are hardly a person, only a body; we only manage to remain a person thanks to the care and maintenance exercised by the anaesthetist. The philosophical story of the nature of humans does not imply that we should believe that part of a human is immaterial.

At this point we have seen something of what Kenny meant, some examples of how philosophy can make meaning "sharper". Now let's see some ways in which language might bewitch our intelligence, as Wittgenstein put it.

Levels of Description

Think again about Saltarella. If people think there is a ladder of explanation what kind of a relation is there between Stories at different levels? For example, there is no doubt that when something happens in the brain atoms and subatomic particles change state. There is always some physical activity in the brain when a perception, thought or a decision occurs. The chemical events in a living tissue are a matter of movements and exchanges of electrons between atoms. Random movements of molecules occur, kicked by thermal energy to produce Brownian Motion that brings them into contact with one another, so that chemical interactions can occur. But we can't identify what is happening in the world seen at our scale when we describe the changes in electrons because all electrons are identical and we cannot tell just by looking at them as electrons which of them is part of the person in whom we are interested. Furthermore many of the atoms of which the body is composed are constantly changing, and even a particular cell often has a life only of the order of days or weeks (although nerve cells often last for many years, sometimes for a lifetime). At what level of description should we describe an event such as a perception, a thought or a decision?

What is physical and what is not?

Here are some things that are physical and have an {x,y,z,t} description:

- A pig
- An apple
- A neuron (nerve cell)
- A nerve impulse
- A photon
- An atom
- A proton

And here are some things that are not physical and don't have an {x,y,x,t} description:

- A belief
- Triangularity
- Bravery
- Ability

Some things we may be uncertain about:

- Radiation (although photons have {x,y,x,t} at least probabilistically)
- Gases (although their atoms have {x,y,x,t} at least probabilistically)
- Energy (although matter, which is interchangeable with energy does have {x,y,x,t})

- Gravity and other physical fields such as magnetism which extend out to infinity.
- A ghost ("white, on the stairs, five foot high, at midnight").

Reductionism and materialism

As science has developed, and as people have become increasingly impressed by how much science can say about the universe, many have thought that the language of science, and particularly the language of physics, is more accurate, more correct, perhaps more truthful, than any other level. They suggest that all other descriptions can be reduced to a description in terms of science, and especially physics. They may say, for example, that the mind is simply the way we experience the physical events, the interactions of atoms and subatomic particles, in the brain. This is a program of *reductionism*. The aim is to reduce all descriptions to those of physics, or at least to languages that are felt to be simpler, nearer to physics, than everyday language. We would expect, eventually, even to be able to explain psychology, including mental events, in terms of physics. People who are *reductionists* are usually also materialists, and believe that whatever life and the events of our lives may feel like to us, there is nothing in the universe that is not physical and material. They put the question, "Why do we ever think that there is anything in the world that is not physical?"

Certainly some things *are* material, for example *a muscle twitch*. We can see one in the large when a limb moves, and measure the electrochemical changes at the microscopic level. On the other hand some things are *not* material, for example *bravery*, which is not a *thing* but a set of abilities (such as the ability to act even when frightened). These latter are not physical *things*, but are not for that reason *non-physical* or *spiritual* things in the sense of *parts of me that are not physical but made of immaterial "stuff"*. Perhaps thoughts, the mind, the will, consciousness, and even the soul are *in this sense* not physical, but need not be ghostly. Remember the advertisement in the fable.

We will come back to this in another chapter. If it is true, then a physicalistic, materialist, reductionist description will be insufficient to capture human nature. For example, we cannot look into the brain and expect to find a lump of *bravery* (although we may well find a part of the brain that is active when a person is brave). For one thing, to be brave requires more than the brain: it requires an environment which contains events which inspire fear and bravery, and so is not describable within the bounds of the individual, but requires a social setting which is threatening. Moreover, it is a person, not part of a brain, who is brave. We give a medal to a person, not to his or her brain (even by proxy). If we misapply language to a problem, we get a silly or inappropriate answer. For example, if you say, "Where is bravery located?" the correct answer is, "On the battlefield", not "In the frontal cortex of the brain." And that is true even though the frontal cortex may be active when someone is being brave. If someone praises me for being brave I may reply, "Oh, go on with you! I wasn't brave. I just did what seemed the right thing to do!" But I don't say, "Oh, don't be silly! I just went along with my brain that was being brave!"

It is not only what happens in the life of the brain that defines what happens in the mental life of a person: after all, you can't struggle desperately to get out of a room if nothing is impeding you[34]. It is the total context, we might say the ecology, in which persons, not just the brain, find themselves that determines the meaning of what is happening. Similarly, there is something a little comical about research "to find the God spot" in the brain, the neuronal centre of religious belief. When people do such research they typically ask people to pray to or think about God, and then look to see what part of the brain "lights up" in a scanner. But what does deliberate prayer by a religious person have in common with, say, a theologian struggling to understand the doctrine of the Christian Trinity, Mother Teresa feeding someone in the slums of Calcutta, or the Archbishop of Canterbury debating the Establishment of religion in the House of Lords? Yet all these are examples of religion.

---

[34] Leaving aside pathological states such as agoraphobia.

Would you expect only the "god spot" to light up in each case? Clearly not. Much more of the brain and indeed the rest of the body must be active to allow these kinds of manifestation of religious belief. How do you think we should speak of causality in such cases? If you conclude from the lighting up of the "god spot" when someone prays that the lighting up is the cause of religious belief, do you really believe that it also accounts for the bishop walking to the House of Lords, and for *all* other activities in his religious life? And suppose the "god spot" does always light up during religious activity, what does it tell you? That's right - that the *person* is religious[35]. A Christian theologian would be right to remark that Jesus came to redeem people, not brains. And of course none of this research tells you whether the belief is untrue or even unreasonable. Remember that if two events occur together, we have to decide which is cause and which effect, or whether the correlation is merely coincidence.

There is much in human nature that cannot be described just by brain events, and such aspects cannot logically be reduced to neuronal activity. Later we shall examine an interesting claim about machines, namely that a machine can be designed in principle to show any behaviour that can be defined in a finite number of words. Can religion, or typically human acts and abilities, be so described? Whether the answer is "yes" or "no" what does that tell us about human nature?

All the questions we have posed in this chapter have been in philosophical, not scientific language. We have shown how philosophy can help us to raise appropriate questions and clarify our ways of thinking about these topics. We see that a philosophical analysis of the language we use to discuss the Fundamental Words can help us to understand how to approach the meaning of such words. Such an analysis is always a very important preliminary step in examining human nature. Only when we have clarified what we mean by the terms we use can we go on to use science to examine what we have defined.

---

[35] Or that all of them involve motivation, and what you have investigated is how motivation is represented in the brain, rather than religion. How can you tell what you have found when the concept being investigated is a rich, multidimensional concept?

Chapter 4

Ways of Thinking: Science

All science is either physics or stamp collecting.

Ernest Rutherford.

A fool. . . is a man who never tried an experiment in his life.

Erasmus Darwin

Believing in Science

Science discovers and explains the physical causes of things. It is fashionable to emphasise Scientific Stories today, but should we give them such a privileged position? Why should we believe scientists when, for example, they say that the universe may have appeared "out of nothing", the earth was created several billion years ago, life arose "by chance", or that thoughts are electrical impulses in the brain?

Although some science was done even in classical Greece, and one can find interesting work in the Middle Ages (particularly in optics and magnetism), science as we know it really began after the Renaissance, and developed rapidly in the 18th century Enlightenment. Science is a wonderfully effective way of getting new knowledge about the properties of the physical universe ("the World") and the laws that govern its behaviour. It is above all a *set of methods* for getting new and reliable knowledge about non-living, living, and psychological entities, in fact the entire universe, everything that exists. If there are aspects of the World that cannot be handled by the scientific method, (and I believe there are,) then they need to be distinguished carefully, because the scientific method is so powerful when it does apply.

Here are typical questions to which without doubt the scientific method can be applied, and to which, until science was used to examine the World, there were no correct answers:

Physics:
What is the shape of the orbit of a planet and why does it have that shape?
What is the nature of light?
What is matter made of?
Chemistry:
What is salt made of?
How does alcohol burn?
What is DNA made of?
Biology:
How do animals and plants reproduce?
How are characteristics of parents passed to offspring?
How does the brain "work"?
Psychology:
How do we see colours?
How do we control our attention?
How do we acquire language?

All these questions can be answered using scientific methods, and we can believe what science has to say about them. No other method is better at answering them. We will have to look closely at what we mean if we claim that what science tells us is *true:* but for now it suffices to say that as scientific research on a topic progresses we approach ever closer and closer to an exact description of the properties of the World.

It is very important to realise that *all science is interlinked*. It is literally true that although they are very different, the scientific claim that the earth began to exist several billion years ago can be linked to the science that allows me to use a mobile phone: the science that allows me to record electrical impulses from the nerve cells in the brain can be distantly related to the Theory of Relativity. To see this let's think about the element *sodium,* whose symbol in chemistry is *Na*.

We encounter Sodium in everyday life in the form of table salt, sodium chloride ($NaCl$[36]). If we throw a handful of salt onto an open fire the flames take on a deep yellow colour. This is the same yellow that we see in some streetlights. In each case it appears because sodium atoms absorb energy from the heat of the fire or the electric current in the lamp and so enter a higher energy state. The atoms emit light when they give up that energy and drop back to their ground state. Our knowledge of atomic physics lets us predict that this will happen, and can account for the particular wavelength (or frequency) of the light that is emitted. If the light is analysed with a spectroscope[37] we see two lines, visible light of only two frequencies, very close together, which we see as yellow, and the wavelength of that light is about 5890.10^{-10} metres[38]. Astronomers find the same lines in the light from the sun, from stars elsewhere in our galaxy (the Milky Way), and even from other galaxies that are trillions on trillions of miles away. So we can conclude that sodium exists in the stars. Measurements of the frequency of light from stars and galaxies let us estimate the size and the age of the universe and identify the chemical composition of the stars. At the other end of the scale, our knowledge of the structure of the sodium atom lets us understand its chemistry, and hence the properties of table salt, how electrical impulses are propagated in the brain, and how to design and operate street lights. Knowledge and theories in all the sciences are unified and interlinked albeit in complicated ways.

In a sense therefore there is only one Science. It does not stand or fall on an isolated fact that may or may not be true, much less on the Authoritative assertion of a single scientist. It is a method of gaining

---

[36] *Cl* is the symbol for the chlorine atom.
[37] A spectroscope breaks up light into its component frequencies or wavelengths, which we experience as different colours.
[38] I shall use this kind of notation throughout the book. The index shows what power of ten the number is multiplied by. So we have $10^2 = 100$; $10^5 = 100\,000$; $10^{-1} = 1/10$; $10^{-5} = 1/100\,000$ and so on. Similarly, $3.4.10^3 = 3400$ and $2.6.10^{-3} = 0.0026$. 5890.10^{-10} is sometimes referred to as 589 nanometers.

new knowledge in which all findings relate one to another. In a sense each aspect of science supports the correctness of all the others. That is one reason why it is so impressive.

The fundamental methods of science

Science is a search for causes and explanations that can be publicly challenged, discussed, and accepted as differences are resolved by appeal to the facts of the universe. There are two fundamental methods used by science, namely *empirical studies* (experiments and observations of the natural world) and *theory building*. Some people distinguish *model building* or *modelling* from *theory building*.

Experiments are systematic ways of causing, observing and measuring events in the World[39], ways that mean that different people can make observations and agree about the characteristics of the *data* that are obtained. If there is disagreement, the experiments can be repeated as often as needed, perhaps with modifications, until people agree on the outcome, and can understand why the disagreements arose. So scientific empirical knowledge, as this kind of Story is called, is *public, agreed by consensus of measurement,* and *self-correcting.* It does not depend on the authority of a single eminent scientist. Empirical studies may consist of observing natural phenomena as they occur (such as astronomical observations on stars, or observing animals in the wild), or of manipulating natural phenomena in the laboratory or the natural environment where what happens is constrained by the investigator who performs an *experiment*.

Theory is the systematic formal description of data or models of aspects of the World. Traditionally theory used mathematics, but nowadays there are other acceptable ways of presenting a theory or model, for example computation, and even verbal, graphical or

---

[39] The physicist Freeman Dyson once entitled a book *Disturbing the Universe*, a phrase which captures the characteristics of experimental science particularly well.

physically embodied models[40]. The purpose of a theory or model is twofold. Firstly, it summarises a large body of data in a form that provides an explanation of the events that gave rise to the data; and secondly, it allows us to predict future events, either exactly or statistically[41]. Often a model provides an efficient cause, a theory a material cause.

Sources of disagreement about scientific knowledge

At any given moment there are many scientific assertions about the World, scientific "facts", which are well-founded, based on evidence and are "certain", although others are far from agreed. There is often little disagreement about experimentally established scientific knowledge, and when there is disagreement it is fairly clear what to do. One can say, "Look, this is how I did the experiment and this is what happened. Didn't you get the same result? How did you do it? Well, let's both do it again. Let's look really closely and see if there is even a slight difference in the way we carried out the experiment that would account for why we don't get the same results." On the other hand, there is great scope for disagreement about theory, and it is usually the over-certain claims of unproven theory that irritate people, whether or not they are scientists. Often there can be several completely different theories for a single set of data. Indeed there can be theory even in the absence of significant data: *string theory* in cosmology is highly developed but has made no empirical predictions to date.

It has been suggested that scientific disagreements can be more profound, that what counts as a good experiment or a clear observation depends on the social context of the investigator or the political system within which science is defined; but the self-checking characteristic of empirical studies can overcome such problems, even if it may take a long time for agreement to emerge.

---

[40] In medicine, for example, if mice are used to test the effects of a new drug prior to its use on humans, we refer to a "mouse model".
[41] A theory or model should always state the boundary conditions within which it is expected to apply and outside of which it is not considered appropriate.

The relation of experiment to theory

A theory is a logical system that is self-consistent and is designed to *describe, explain,* or *model,* data. Data are observations about events in the world. They are an account of what happens. The difference, more or less, is that an explanation tells you *why* something happens, in particular why the data take the form they do. It leaves you with a feeling of, "Ah! *That* is why that happens!" A model may not have the force of an explanation, but at least it will *predict* the form the data will take. For example, a complex mathematical equation may predict the height of tides day by day reproducing the sequence of numbers that represent the daily tidal heights, but not suggesting an underlying mechanism. On the other hand one can use the relative masses and orbits of the moon, sun, and earth and the theory of gravity to give a causal account of why tides rise to the height they do. Ideally one hopes to have both a model and an explanation: we can show that the way in which the sun and moon affect the earth gravitationally is just what one would expect to produce the mathematical form of the tides. This is what Newton did with his theory of gravity.

To take another example, observations show that planets appear to move in complicated orbits. If you plot the position of the planet Mars, night by night, with respect to the "fixed" stars in the night sky, it appears to move sometimes from left to right, and at other times from right to left, sometimes from higher to lower in the sky, and sometimes from lower to higher. The variations of the position of Mars relative to fixed stars had been accurately known for many years. Those observations are the *data* to be explained or modelled. The data in this case are a series of numbers, representing the position of the planet at different times. Suppose that someone offers a theory about the movement of Mars. I will accept the model as accurate if it tells me at what position to find Mars each night. The more often the model correctly predicts the position of Mars, the more firmly I believe in the model. The fact that science can make predictions like this is one very powerful reason why many of us "believe in science". From time to time someone may refine the model. By slightly changing the predictive equations they can make the prediction of Mars's position more accurate; or instead of saying where the planet is at 24-hour intervals, they can predict it second by second. Predictions using this kind of model are so accurate that probes have been sent from earth to intercept comets and asteroids many millions of miles away in deep space. At the time of writing the Rosetta space probe has just landed on a comet after travelling through space for nearly ten years to its rendezvous.

It is sometimes said that a good theory is only changed by a better theory, not just by facts that disagree with the theory. One tends to stick with a theory even when it is known sometimes to make incorrect predictions, if one doesn't have an alternative theory. Every now and again a really radical change may be proposed. For example the system of Newtonian equations used to predict the position of the planet Mercury were replaced by a new set of equations when Einstein's theory was able to predict changes in Mercury's orbit that Newton's equations could not account for. For many centuries most people thought that the planets revolved around the earth, and that the latter was at the centre of the Universe[42]. But even once the Copernican model had been proposed, namely that all the planets including Earth revolved around the sun, there was no good explanation of what caused their movements.

The original equations used to predict planetary positions used equations for circular motion. These were replaced by equations representing elliptical motion (Kepler), and later still by a theory that explained why the motion should be elliptical (Newton)[43]. In more modern days Einstein's equations representing the geometry of space-time have superseded those of simple elliptical motion. Now we say the equations represent not forces acting across space, but distortions of the structure of space-time caused by matter, in this case the huge mass of the sun. The more accurate the prediction, the better the model; and *vice versa*. One reason that it is so hard to make money on the stock market is that even if it is possible to fit

---

[42] One of the earliest pre-scientific theories of the cosmos proposed in about 500 BC was heliocentric.

[43] When navigating a yacht by using a sextant we still assume that the sun goes round the earth. The mathematics are much simpler if we do, and the critical measure is the apparent position of the sun with respect to the horizon, given the time of day and the date.

mathematical equations to stock market fluctuations, at best such equations model but do not explain the form of the fluctuations. Climatology is another case where the measurements are precise, and some of the underlying theory well understood, but the immensely complex interactions among a large number of variables makes prediction difficult.

As a biological example, consider Mendel's laws of genetics. He collected data on the way in which characteristics of pea plants passed from generation to generation. Pea plants occurred as tall or as short plants, and seeds were rough or smooth. His model predicted that when two plants are cross-fertilised, the characteristics of the offspring would be those of one of the parents, rather than blending the characteristics of both. If the offspring were then interbred, in the next generation the characteristics of their parents would appear in the ratio of 3 to 1. If a tall plant is crossed with another tall plant, the offspring may be tall or short, and Mendel's model predicted what proportion of each would result. In more complicated cases, as we shall see in a later chapter, the characteristics do not sort themselves out so clearly, but the underlying mechanism can be shown to be the same or similar. Although he could make accurate predictions, Mendel did not understand *why* the model made correct predictions, *why* plants' characteristics appeared in certain proportions. The existence and properties of genes were unknown. He had a model, but not a theoretical explanation, an efficient but not a material causal account.

To answer "*why*" questions we need *explanation*. Here again, the strength of support for and power of the explanation is part of the reason that we "believe in science". In particular, the *principle of falsification* is central to modern science. There is a logical problem about proving a theory to be true. For example, in the past every time I have boiled water (at normal atmospheric pressure) it boiled at 100 degrees Celsius (100°C). But there seems to be no logical reason to say that because such data appeared in the past that guarantees that the next time we boil water it *must* boil again at 100°C. Data alone cannot *confirm* an explanation. A single counter-instance is sufficient to disprove the explanation, since the explanation is meant to apply to *all* cases. *This is one hallmark of a scientific explanation.* There may be no evidence[44] until now to suggest that an explanation is wrong. But someone who proposes a scientific explanation (theory) or a model *must specify some prediction or predictions that, if fulfilled, will prove the theory wrong* if it is to count as science. In practice any scientist who proposes a theory believes it to be true, or at least that it is a good working approximation to the truth. Such a scientist should however be prepared to give up his theory if data show that it cannot predict what will occur[45]. Science proceeds by self-correction. Even if a scientist is so wedded to his theory that he refuses to see weaknesses in it, others will certainly perform the necessary reforms, although it may take time. As Darwin[46] wrote,

> False facts are highly injurious to the progress of science, for they often long endure; but false views, if supported by some evidence, do little harm, as every one takes a salutary pleasure in proving their falseness; and when this is done, one path towards error is closed and the road to truth is often at the same time opened.

It may not be possible at a particular moment actually to carry out the crucial experiment: but it must be possible *in principle* to disprove the theory. Only such theories are scientific. For example there may have been someone in 1400 AD who said that he believed that there are mountains on the moon. Now in 1400 there was no direct observation possible to refute that idea: telescopes did not yet exist. But it could still have been a scientific theory, since in principle it could be refuted: "If ever we can somehow look much more closely at the moon, or go there, we will find there are mountains, even though from here it looks smooth to the naked eye." Theory in subatomic

---

[44] Although that is commonly accepted, there is a flaw in the argument; but I won't go into it here. We will discuss it in the chapter on Probability.

[45] Although this is the hallmark of science, it is a necessary attitude in discussions of all kinds. If you are having an argument with someone, there is no point in continuing unless you both are prepared to say what evidence will convince you that you are wrong. If one or both of you say that whatever evidence or argument is made you will never change your mind, it isn't worth continuing the argument (except to pass the time!).

[46] C. Darwin, (1871) Descent of Man. p385.

physics predicted for many years the existence of a particle called the "Higgs boson"; but the theory also predicted that the amount of energy needed to find it was greater than we could then produce. Recently the Large Hadron Collider at CERN, has produced enough energy to generate the Higgs boson, and it now seems to have been found. This is an excellent example of where theory developed in advance of data. But notice that the theory predicted something that could be measured *in principle* when time, money and technical expertise allowed.[47]

Let's look at how these ideas are used in practice. We might agree that the data describing the motions of the planets show that they move in elliptical orbits around the sun, and that the mathematical model of elliptical orbits is accurate enough to be convincing. We have then a scientific model. But *why* do planets move in *elliptical* orbits? Well, here are two possible explanations. One is that invisible spirits, perhaps angels, push them around by beating their wings. An alternative is that there is a force, gravity, that is produced by any piece of matter, so that all pieces of matter attract one another, and the force of gravity is proportional to the product of the masses of the planets and inversely proportional to the square of their distance apart. The first of these would not be a scientific explanation even if it were true, because there is no way to test for the presence of angels. Their existence and "propelling function" cannot be falsified in any way. The second explanation is scientific, because we describe experiments that will provide data that relates force to mass and can measure the force of gravity between many objects. The resulting data show that the forces between objects do indeed depend on their masses, and gravity does indeed vary with the square of distance, and from this it follows (via a set of mathematical equations) that stable orbits will have to be elliptical[48]. When Newton wrote his *Principia* there was great resistance to his ideas by some scientists because gravity was a force that operated at a distance without contact

---

[47] Note that even if the new particle turns out not to be *the* Higgs, this example is still valid as an illustration of scientific method.
[48] It even predicts that if the universe had either only 2 or as many as 4 or more spatial dimensions the orbits would be circular, and unstable. We can take it that those predictions are not *directly* testable!

between bodies, and such an idea was very difficult to accept at that time. After Newton it was accepted because it allows predictive equations to be solved. Note that Newton did not know why gravity exists. Nor do we. At present it has the status of a "brute fact" of the universe. Although Einstein's theory of the distortion of space-time by matter pushes the explanation back one level, at present his theory also is a brute fact.

Let me emphasise again, it is *because the predictions of scientific theories are fulfilled that I believe in the findings of science and in science as a way of gaining knowledge.* They are not *always* fulfilled, but they are fulfilled in an extremely impressive proportion of occasions; and when they are not fulfilled one can see what to do next in order to find out why the prediction failed. I also believe in science because it has a built-in way to correct any incorrect theories, namely empirical data collection. Consider Quantum Theory. Many of its predictions seem completely bizarre if all one knows is the everyday world.[49] But it has existed for about a century, and *every single prediction that it has made in that time has been found to be correct.* That includes even completely non-intuitive predictions such as the ability of something to be in two places at once! Furthermore, the same theory has supported the design and manufacture of a huge range of devices and products that are ubiquitous in our lives, including modern computers, mobile phones, and so on. That seems to me as good a reason as one could possibly have for believing in anything, even if I do not fully understand the theory, and cannot myself do the mathematics or experiments needed to make predictions. *Science generates its own authority by its success in prediction and explanation.* Furthermore, I am more and more willing to believe its predictions in new fields of knowledge because of its great success in the past, even if I am not initially predisposed to believe those predictions - providing only that I am sure that the domain of knowledge is such as to place it logically within the domain of science.

---

[49] See R. Feynman. 1990. *Q.E.D. The Strange Theory of Light and Matter.* Princeton University Press.

There are some special cases where predictability is limited, such as *complexity theory*. This is a relatively new branch of science where we cannot reduce phenomena to simple causally interacting components, but where the aspects of the World are very richly interactive and often have a very large number of parts. Examples are the weather, brains, and even societies. It is now known that such systems have properties that make it almost impossible to make detailed predictions about outcomes. They are very sensitive to the starting conditions when the prediction is made, and there may be several possible outcomes from what seem to be the same starting point. Furthermore, in such systems completely new properties may appear spontaneously. The systems are said to be "self-organizing", and our understanding of the theories that underlie the phenomena is steadily progressing. At least we understand why prediction is difficult. Despite these difficulties science remains the best way to acquire new knowledge. In the case of complex systems any other approach is worse.

The relation of science to other Stories.

Are there some kinds of knowledge that are *not* scientific? Certainly literature and history are not scientific although enormously valuable as human achievements. Philosophy is not science; and I do not believe that classical economics is a science. Some parts of psychology are not science. To have a good scientific theory you need good data, and to get good data you have to be able to define what you are measuring. An idea that has been around for a long time is that of psychological *forces* that make people do things, such as show neurotic behaviour. Intuitively the idea of a psychological *force* seems attractive. Many people have felt at some time compelled by depression to stay in bed, for example. But the problem that stops this kind of psychology from being a true scientific account is that things are missing. *Force* in physics, where the notion originates, is defined by an equation, $F = ma$, where F is the force, m is the inertial mass of a body, and a is its acceleration. (You have to push hard to make something massive accelerate quickly.) But in psychological theory there is nothing that corresponds to *mass* and nothing that corresponds to *acceleration*. So the concept of a psychological force, whatever it may mean, cannot be the kind of force that pushes physical objects. So it cannot be a cause in the way that physical forces are causes. Perhaps we should regard it as a fable or metaphor[50].

Think also about the currently fashionable research on "happiness". The definition of happiness in such research is often so imprecise that you cannot be sure what is being measured or even whether you are measuring it. You may feel the same about love, or religion, or many other topics in social and psychological "science". It is generally agreed nowadays that Freudian psychoanalysis is not a scientific theory, because almost none of its predictions are falsifiable. If I want to generate a science of religious belief I must be able to define it in such a way that I can collect data. But there is very little agreement as to what counts as religious belief. That makes it difficult to see how we can have a classical science of religious belief, and also has direct relevance for neuroscientific research that tries to identify "the God spot" in the brain using fMRI.[51]

In the case of psychology however there are many cases where no methodological problems arise to prevent scientific predictions. One example is my own research on one's ability to divide attention. I played two auditory messages to listeners simultaneously, and asked them to detect faint signals in one or both. I measured the probability that they could detect the signals, and converted the measures into two statistics that are also used in engineering and decision theory. I could explain the behavioural phenomena on the assumption that attention can be modelled as a "volume control" plus a mechanism that makes people more or less willing to accept evidence as a function of its probability and quantitative value to them. The numerical predictions from this model were fulfilled. Another of my experiments has been repeated by two other researchers independently at 20 year intervals,

---

[50] The same can be said of the concept of "energy", *chi*, in traditional Chinese medicine,
[51] Similar problems apply to many neuroscience projects. There is nothing special in this regard about religion.

with identical quantitative results[52]. So in some cases even "mental" events can be modeled scientifically.

There are some problems in physics, most notoriously in cosmology, where it is unclear what, if anything, modern theory predicts, and how it can be falsified. As mentioned earlier String Theory, which has been proposed as the long-sought "theory of everything" has been criticised because while it is self-consistent mathematically, and hence a usable *model*, it has made no falsifiable predictions. If that is so, then it is *not* a scientific theory even if it is a beautiful mathematical theory[53].

This is a good place to recall that it is a task of philosophy to make concepts sufficiently precise for them to be used by science. As Kenny said, "It sharpens concepts". It is important to understand the role of definitions in science. Sometimes, particularly when looking up meanngs in a dictionary, people think that a definition gives the absolute meaning, the Platonic form or essence of a concept. That is not how we use definitions in science. We do not look for the essence of a concept by defining it. Rather we work with *operational* definitions. We define a concept in terms that make it possible to investigate it. For example, in my experiments on attention, I did not ask what attention really is. Rather I defined attention as what is measured if I ask someone to listen to two messages at once and measure the probability of their being able to hear two messages at the same time, or to select one and reject the other. This gives me a working definition that is clearly related by "family resemblance" to the concept as used in Everyday language. Science can work with operational definitions, but must show how they relate to their Everyday "family" to justify their use.

---

[52] I put emphasis here on psychology because of the intimate relation of psychological research to the Fundamental Words. See Moray, N., "Attention in Dichotic Listening: Affective Cues and the Influence of Instructions," *Quarterly Journal of Experimental Psychology*, 11, 56-60, 1959. Moray, N., M. Fitter, D. Ostry, D. Favreau, and V. Nagy, "Attention to Pure Tones," *Quarterly Journal of Experimental Psychology*, 28, 271-283, 1976.

[53] For an interesting discussion of this point in subatomic physics, see Veltman and Veltman, 2003.

So, to summarise, we believe in scientific explanation and scientific knowledge because, in many cases science makes accurate predictions, the research is repeatable, and because it is falsifiable – it tells me how to test it. Science is inherently self-correcting and explanatory.

More about causes

Science adds to our knowledge by revealing causes of phenomena. To know the cause of something is to have an explanation for it. We saw in a previous chapter that there are several kinds of causes, and not all of them are what science has traditionally considered. Science gives us *efficient* and *material* causes for natural phenomena. Often that is enough to satisfy us, to make us feel, "Aha! Now I see!" Often too, as we saw, this feels as though we are being led inexorably to "deeper" explanations: muscles explain movements, biochemistry explains muscle action, and physics explains biochemistry. But we also saw that we should not accept the idea that ever deeper causal accounts always give us ever more powerful explanations. Quantum physics simply does not explain why a high jumper attempts a record leap. To say that the cause of the jump was the athlete's desire to win is a simpler and more satisfactory explanation than to say that somehow there was a simultaneous occurrence of a very large number of quantum changes of state in the atoms of which the muscle is composed. But if there is always going to be this kind of choice available as to the level of explanation, is there any way to decide what level we should use for a particular explanation? What kind of cause goes with what kind of question? What kind is best suited to explain human nature?

There are two traditional guides on to how to choose a level of explanation. The older dates from the 14th century, and is called *Ockham's Razor* after the philosopher William of Ockham. He proposed that one should always look for a simple explanation and not postulate unnecessary entities when explaining phenomena. Keep it simple. If we find that how a plant grows toward the light appears to be completely predicted by the effect of light on a chemical in the growth-point of the plant, we should accept that as the cause and explanation. We should not go on to say that there must be a special "élan vital" characteristic of living tissue that urges it to grow towards the light. Ockham's Razor excludes magical and supernatural explanations where we have adequate scientific evidence and theory.

The second guide is more modern, and was proposed by a 20th century psychologist. It is known as *Lloyd Morgan's Canon*. Morgan proposed that if we can account for events in the biography of an organism (human or otherwise) by a low-level explanation, a more complex explanation should not be proposed. A particularly important example is language. There is a tendency to think that non-human animals have a language in which they talk to themselves and one another in the way that humans do. Do they have minds? Not, I think, in the sense that humans have them; and the reason is not that only humans have GIMs, since they don't. Rather it is because of their special capacities, abilities, that we use the word *mental* to describe human biographies. What we refer to is often the human ability to use symbols, to have the kind of language that we have, and this is missing from all other animals.

Bees have a language that allows them to communicate the direction and distance of a food source to other bees in the hive by dancing. Bird song is a kind of language, as are the various barks, growls and gestures of dogs. But these languages are very constrained. Apes and monkeys, and even cetacea (whales) may have quite extensive vocabularies. In some cases animals are born with their language almost complete and the range of "statements" that they can make are limited to a closed set to which they do not add new words. In others their languages are to some extent modifiable by learning. Some birds seem to be born with a theme song typical of their species but which can be modified by hearing the song of birds among which they grow up: and of course parrots can, indeed, *parrot* even human speech sounds. But none of these have open-ended languages like human language. Bees can talk about a great variety of distances and directions, but they cannot talk about social reform. Birds don't have conversations about fashion or politics. Lloyd Morgan's canon asserts that we must not say, "How do you know? Perhaps you just don't understand them." Humans, with a small set of sounds (phonemes) or written symbols, can generate an infinite number of sentences of unlimited grammatical complexity, many of which have never been uttered before. There is absolutely no evidence for this in other animals, so we have to reject the possibility. No animals, not even the great apes or whales, have such language. And it is precisely the ability to manipulate symbols, to think using symbols that we have invented rather than inherited, that makes us need a special word to describe human ability, and that word is *mind*.

It is worth thinking about this a bit further. I don't want to insist that the only way a person can think is by using symbolic language. Perhaps one could think by manipulating images in the mind's eye, so to speak. A chess player does not work out logically all possible moves using language or logic, but often uses pattern recognition. There are very strong reasons to think that this is how good players, particularly Grandmasters, think. We can imagine someone looking at the board, and then, in his or her mind's eye, imagine the pieces moving around the image, and looking at the image to see what its properties would be if such a piece were to be moved in such a way. Or I can imagine a street in a town where I used to live, and I can work out how many shops there are between two named shops by looking at the image and counting the shops. (I once answered a question in a pharmacology examination by calling to mind the page where there was a diagram of the molecule of a particular drug and then looking in my mind at the arrangement of the atoms in the molecule.) I don't see why we can't call image manipulation in our imagination a way of thinking, even taking Lloyd Morgan's view. But the typically human, characteristically human way of thinking, the way of thinking that marks us off from any other species we know, is the use of symbolic language with complex recursive grammar and syntax, and that is central to what we mean by using our mind.

Close inspection shows no behaviour in other animals that cannot be explained without such a language. Therefore we should not propose that a particular piece of behaviour, such as a dog whining at the door when expecting that his master is coming home soon, means that the dog must be saying to itself, "Oh, if only he would come! He will take me for a walk!" If there is an alternative explanation in terms of learning (conditioning) or instinct without language, we should adopt that explanation rather than the higher-level explanation.

A particularly elegant example of the application of Lloyd Morgan's Canon (and indeed also Ockham's Razor) is the explanation of why certain insects always gather in dark places, which was mentioned earlier. Although it was initially thought that they actively seek out low light levels ("negative phototropism"), it turns out that there is a simpler explanation. When bright light falls on them, they move in random directions. When little or no light falls on them, they do not move. Hence as time passes more and more of them must inevitably end up in the shade. No "negative phototropism" or "dark-seeking instinct" is needed to explain the behaviour, just random movement in the light and no movement in the dark. A material cause is sufficient, and there is no need to invoke a final cause.

Neither Ockham's Razor nor Lloyd Morgan's Canon are themselves justified by scientific evidence. They are adopted because they make science work consistently and effectively. They are axioms of science, not conclusions from evidence[54]. They are aspects of philosophy applied to science. They should make us very wary indeed of adopting explanations of human abilities in terms of "spiritual", or "non-physical" properties. We should ask whether the latter are really needed, or whether Scientific Stories are sufficient, at least in principle. To adopt these axioms means that we must look carefully at what kind of explanation we need for the Fundamental Words. Why do we ever feel that mental activities *force* us to think in terms of my possessing a non-physical *part* or *component?* Can we not find a simpler way in which to talk?

It may be helpful to see how we could apply scientific method to research on a phenomenon that many people feel needs a non-material explanation. How would one study telepathy, or as it is sometimes called, *extrasensory perception* (*ESP*)? There are two questions to think about. The first is how one would perform experiments to determine whether people can indeed communicate "from one mind directly to another". What operational definition should we adopt?

---

[54] Compare this with how the Logical Positivists defined meaningful statements. They said that only statements that were scientifically verifiable were meaningful. But that statement is itself not verifiable, and so must be meaningless.

The second is, supposing that we were to find that the phenomenon actually exists, how should we explain it?

Science and ESP

People sometimes claim in Everyday Stories that they know what another person is thinking. Furthermore, they say that this shows that one mind can communicate directly with another mind, without any physical connection between them. People tell stories about how they "knew" when someone dear to them died, even though the person was many miles away; or they note that just as they were going to say something to another person that person made the same suggestion: "It was just as if we were sharing the same thoughts." Now the ability of people to communicate without any physical connection between them would be extremely interesting if it could be proved to happen. How could we set about determining the truth scientifically?

There have been many attempts to check whether claims about knowing that someone had died or suffered an accident when they were far away are true. It has to be said that the evidence is unconvincing. Almost without exception it turns out that people cannot establish exactly when they had the thought about the other person. Was it at exactly the moment when the accident occurred? Or shortly afterwards? Or shortly before? Indeed, what counts as "at the same time"? The same day? The same minute? The same second? The same millisecond? Or just the same week? It is very difficult, given the fallibility of human memory, to establish the facts, and there is no really convincing evidence of these events. Moreover, if such a synchrony could be established, how do we know that it is not just a coincidence? How often have we had the thought that someone has had an accident when no accident happened? And if someone says that they can't remember many cases when they incorrectly had such a thought, how can we be sure that they have not just forgotten them? Perhaps we forget unsuccessful thoughts more readily than successes.

Since the existence of non-physical communication, ESP, would be so extraordinary, we would need extremely strong evidence for its occurrence if we were to believe in it. (As a rule, we should ask for exceptionally strong evidence for very unlikely events, and be satisfied by relatively weak evidence if something is highly probable.) Moreover, there is strong evidence that people are easily deceived, indeed deceive themselves, about things they would like to believe. For a particularly telling example, see *The Road to En-dor*[55], already mentioned. In that true account of people who pretended to send spirit messages, some of those who received the fake messages refused afterwards to accept the explanation that it had been a hoax, and insisted that it must have been ESP by spiritual communication.

So if anecdotes are not convincing, how should we investigate the possibility of ESP? Most of the scientific work on it has used a technique something like the following. We start with an operational definition of what would count as two people communicating directly mind-to-mind. Two people are chosen as experimental subjects, and sit in rooms separated so they can neither see nor hear one another. One, whom we will call the "sender", S, tries to "send thoughts" to the other, the "receiver", R, who tries to receive them. To make the measurement of what happens as objective as possible, S has a deck of cards each of which has a different abstract pattern on it. One such set are the so-called *Zener cards* that were used in a long series of experiments at Duke University. There are five different Zener card patterns: a hollow circle, a Greek cross, three vertical wavy lines, a hollow square, and a hollow five-pointed star. There are 25 cards in a pack, five of each design. The two people know the exact time using synchronised clocks, and at particular moments, say once every 30 seconds, S looks at the next card in his shuffled pack, and tries to sends his thought to R. The latter writes down what she thinks was the pattern that was sent at that particular moment.

How can we tell whether any information has been transmitted? If R is just guessing, and really receives no information from S, then the "hit rate", the proportion of answers that coincide with what is "sent", will be about 0.2: about 20% of the answers will be correct. There are five cards, so if R guesses which one is being sent without

---

[55] E. H. Jones. *The Road to En-dor.* 1942. London: Bodley Head.

any information, R has a one in 5 chance, 20%, of guessing correctly. Because of random chance fluctuations we would not expect exactly 20% correct on each trial, but the number of hits should fluctuate around 20% ± x, ("20 plus or minus x"), where x is a small number of the order of a few per cent. In general, that is exactly what happens in almost all these experiments. However, once in a while, R has is a run of correct answers. Now, suppose that R gets 3 in succession correct. What is the probability of that occurring? (In the next chapter we will see even more clearly how important is probability to the conduct of science in genereal.)

The probability, p, of correctly guessing one card is $p = 0.2$ (i.e. 1 in 5, or 20%). If we assume that each guess is independent of the others, then the probability of two correct guesses is $p = 0.2 \times 0.2 = 0.04$ (i.e. 1 in 25): and the probability of getting three correct in succession is $p = 0.2 \times 0.2 \times 0.2 = 0.008$ (i.e. less than 1%). Is that enough to convince us that there really has been a transmission of information from S to R?

In much behavioural research scientists say that the results of an experiment are not due to chance if the probability of the outcome being due to luck is less than 1 in 20, ($p < 0.05$)[56], and will almost always accept a result if the probability that it is a chance effect is less than 1%[57]. But nonetheless there is great resistance to accepting that a run of three correct in a Zener card experiment is evidence of transmission. Occasionally much longer runs have been reported, with probabilities approaching $p < 0.000001$. Yet even so, the results have not been accepted by the scientific community. Why not?

The first reason is, as we saw earlier, that the claim is that information has been transmitted by some non-physical channel of communication, one mind communicating with another. That is such an extraordinary

---

[56] The symbol < means "is less than".
[57] Experimental physics usually requires probabilities of 1 in 1 000 000 or smaller. Where possible we should ideally require similar probabilities in other disciplines, but it is often not possible because of the amount of variability and the cost of research. In general, to double the accuracy you need four times as much data.

claim that even a very, very low probability of the result being due to chance is not enough to convince many people. Furthermore, while we have data, we have no theory or explanation. And there are two more important objections. Suppose that there really is some transmission of information, why should we conclude that ESP uses a "spiritual", "mental", non-physical channel of communication? If communication takes place by no known physical means, the simplest conclusion is that is taking place by an unknown *physical* means. To make the jump to an *unknown unphysical* means is just gratuitous. What we should do is to try to find the unknown physical communication channel. It might be that the electrical fields of the neurons in the sender's brain act as transmitters, and the signal can be picked up by the receiver's brain acting as an antenna[58]. So if we really think that the results cannot be due to chance, the next step would be to look for a model of the physical basis of the ability and develop a theory of how transmission occurs.

Here we run up against the second objection. These experimental results are not reliable. The long correct runs occur rarely, at apparently random times, and even with a particular sender and receiver are not usually repeatable. But if communication were really occurring we should be able to repeat the experiment; and *only* if the experiment is repeatable does it count as good science. There is another way of analysing the results, using not just direct probability calculations but what is called the *mathematical theory of communication*[59] which is more theoretically appropriate, and when it is applied shows that even where the long runs of successes occur there is statistically no significant transmission of information.

---

[58] There are recent experiments where wires have been implanted in the brain of an animal and by processing the signals with computers researchers have transmitted signals directly to the brain of another animal. http://telegraph 28/2/2013. This is of course a quite different phenomenon. Similar results have been reported recently by a group working with humans. See *Conscious brain to brain communication* by C. Graus and others in PLOS ONE, 19 August 2014. 10.1371/journal.pone.0105225.

[59] C.Shannon, and W.Weaver. 1947. *The mathematical theory of communication.* Urbana: University of Illinois Press.

So we don't accept the truth of ESP research as a Scientific Story. But notice that it makes clear how Scientific Stories are typically created. We choose a phenomenon to investigate. We develop an operational definition of the phenomenon. We decide on a criterion for telling whether the results of the research could be just chance. We carry out the research and compare the results with the predictions of some mathematical or other model. We try to repeat the research to show that it is a reliable story. And finally, we put the results into a theoretical framework. Sometimes we review many past studies and combine them quantitatively using what is called *meta-analysis*. It is because we have an agreed set of conventions for doing this that Scientific Stories are so powerful, both as explanations of why things happen, and as sources of new knowledge. The methodology protects us from accepting chance events as meaningful. So an understanding of chance and probability is central to understanding how science works. Knowing this we can better see what modern science can contribute to understand human nature, but it also is essential to understand how probability affects our knowledge of the natural world.

Chapter 5

Probability

Probability is the very guide of life.

Rev. J.Butler. *The Analogy of Religion.* 1736.

To understand the claims of science we must understand something about probability, because science and scientific method depend on it. So this is a chapter about using mathematics to understand the World. It is sad that many people fear mathematics and even take a perverse pleasure in boasting that they can't use it. The fact remains that all empirical scientific knowledge, (that is knowledge obtained from experiments and observations of the natural world,) is probabilistic, so we need to understand something of the underlying mathematics. How can the claims of science be certain if it is founded on probability?

To understand how intelligence is measured, to model the birth and death of stars, even to know whether to expect Saltarella to do better on her next jump involves probability. When a scientist claims that what she has predicted has in fact occurred she asserts that the outcome of the experiment was caused by a process or mechanism described in her model or theory: it was not due to causeless chance, or to one or two causes chosen at random, but to a cause specified by the theory. Be assured that the mathematics is not very demanding. Treat it as a kind of mental jogging to keep the mind fit!

Our notions of probability emerged in connection with gambling in the 17th and 18th century[60]. Professional gamblers wanted to know how much they should pay to take part in games of chance, and put the question to some of the great mathematicians of the day, such as

---

[60] J. Bernstein. 1998. *Against the Gods.* John Wiley & Sons.

Fermat, De Moivre, and Bernoulli. Today probability has become a fundamental way of looking at the world, and should be a primary topic in education.

There are several interpretations of what it means to say that an event is *probable*, or due to *chance*. Firstly, there may be strictly no possible causal account of why something happens: it is just a truly chance event, a stochastic[61] event, a probabilistic event. Events in quantum physics are of this kind. A certain proportion of atoms of a radioactive element will disintegrate within any stated period, but there is no way that we can say which particular atoms will disintegrate. Which ones disintegrate is the result of a stochastic process, a matter of chance, a matter of events matching numbers that represent it as a probability. There is strictly speaking *no* cause for the events other than the necessity to match the mathematics of probability. We might call this *absolute empirical probability*.

The relation between randomness, chance and cause is subtle. In the case of quantum uncertainty to say that events are random is to say there is no cause for them, but that is not usually true of everyday events that are probable. If I toss a coin to decide who shall start a game, I imply that I cannot determine which side of the coin will be uppermost when it lands. But certainly, in some sense, which side comes uppermost depends causally on, among other things, the way I flip it. The probability of one of a set of possible events is equal to 1 divided by n, where n is the number of possible outcomes. Since a coin has two sides, $n = 2$ and we say that the probability that it will land so as to show us "heads" is one-half, ½, that is $p = 0.5$ where p is the probability of the event. Over a long run of tosses we expect to see more or less 50% heads, although it will be very rare to see exactly 50%. Usually the observed probability of heads will be a little more than $p = 0.5$ or a little less, and occasionally the value will be surprisingly far from $p = 0.5$. Similarly, if we toss a die many times, $n = 6$ and we expect each face to come uppermost with a probability of about one-sixth, a proportion of about $p = 0.167$. If we spin a roulette

---

[61] From the Greek στοχος meaning *aim*, and referring to the scatter of hits around the centre of a target.

wheel we expect the ball to fall into the red on about 50% of trials, to land on one of the numbers 1 – 9 with about $p = 0.25$ (one-quarter, 25%), and to land on any *one* particular number with a probability of just less than $p = 0.028$. The probability of drawing a face card (King, Queen, Jack) at random from a pack is about $p = 0.23$, while the probability of drawing a non-face card is about $p = 0.77$.

This way of presenting probability is said to interpret probability as the *relative frequency* of events. We cannot predict, on any one trial, exactly what the outcome will be ("heads", "5", "red", "23"); but although we do not have exact knowledge about how the physical causes determine the outcome, we can say what, on average, to expect. To say that such events are probable is not to deny they are caused. There is undoubtedly some set of physical influences that when taken together is the cause of the outcome. But there is no way we can follow the unfolding of the causal history: it is too complicated and unfolds too rapidly, and is difficult or impractical to observe. The process is not necessarily complicated, but it is complex in the sense of complexity theory that we mentioned in an earlier chapter. So outcomes are uncertain, and must be expressed only as the relative frequency of the different possible events. Even a biased coin or a loaded die will not always give the biased outcome. The outcomes will not be absolutely predictable on a trial by trial basis.

We make some important assumptions when interpreting events as probabilities. Where we expect equally probable outcomes we assume that the coin is unbiased, and that it will never land balanced on its edge. We assume that the die is not "loaded", that the surface on which it is thrown is smooth and hard, so that it must end with one face uppermost, and that it is thrown fairly. We assume that the roulette wheel is "fair", and that there is only one "0"[62]. It just might

---

[62] In some casinos there are two slots labeled "0", which increases the odds against the player and in favour of the casino very slightly. With only one zero the probability of any particular number, say 17, coming up is approximately $p = 0.0270$: with two zeros the probability is approximately $p = 0.0263$. The probability of red coming up is in the two cases approximately $p = 0.486$ and $p = 0.474$. Over thousands of trials this makes a substantial extra profit for the casino.

be possible in principle, if we had access to all the initial settings of the game, to the equations of motion and the equations of mechanics, to calculate how many times the penny would spin before it hit the ground (if only we knew also all about the thumb of the person tossing it, its force, its elasticity, etc.). But in practice we treat the event as if the physical causal events are unknowable, and hence say that the outcome is a matter of chance, and is probable, has a probability of p, and so on. We should really speak about probabilities as proportions; the probability of an event is represented by a number never greater than 1 (certain) and never less than 0 (impossible). But often people interpret proportions as percentages and say that "there is a 50% probability" when what they mean strictly speaking is $p = 0.5$.

Probability enters science in several ways, all reflecting uncertainty. There can be uncertainty of knowledge, uncertainty of measurement, and finally absolute empirical uncertainty. When a weather forecaster says, "There is a probability of 33% of rain tomorrow", he cannot be more accurate because some knowledge is missing, and because of the extreme complexity of the physics of the atmosphere, so that the necessary calculations cannot be carried out to make the prediction more accurate. This is an example of uncertain knowledge. Obviously this probabilistic prediction is not really a "relative proportion" probability, since we have not experienced a large number of tomorrows on which we can base the estimate. (For *today* there is only one *tomorrow*, and it has not yet occurred!) Perhaps it is just the expression of how much one would be prepared to bet on rain tomorrow. Or does it mean that definitely one-third of your garden will get rained on? It is left as an exercise for the reader to think of other interpretations.

A different source of uncertainty in science is uncertainty of measurement. This problem is most severe in the social and behavioural sciences. That is not because social sciences are about humans rather than inanimate objects, but simply because what is measured is variable even when repeated measures are made on the same person. The same problem would arise in physics if the charge on the electron varied substantially in a random way. In experimental physics it is not uncommon to measure to an accuracy of better than one part in a million, in some cases to one part in 1000 million. (The magnetic moment of the subatomic particle called the muon is known to 9 places of decimals, one part in a thousand million.) Repeated measurements of an entity give almost exactly the same value. We can make measuring instruments that are as accurate as we need to measure the phenomena. In measuring the dimensions of the components of living cells we can easily measure to an accuracy of the order of the wavelength of light, between 100 and 1000 nanometres[63]; and using devices such as the Atomic Force Microscope we can measure the structure of cells to accuracies comparable to the size of single atoms[64].

In measuring animal behaviour we time events to fractions of a second, and in experimental psychology we often measure times to an accuracy of milliseconds[65]. But unfortunately in many measurements of human biography the variability in the data is of the same order of magnitude as what we measure. We cannot say, for example, that the time it takes a person to respond to a light is exactly 180 msecs. Instead we have to give an average such as the *mean*, μ, which is a best estimate based on a large number of measurements. With any such measurement we should also provide a measure of how variable were the measurements[66]. We get the mean by adding up the values of all the measurements we have taken, and dividing by the number of measurements (data points). We usually measure the variability using the *standard deviation*[67] (sd, σ) or its value squared, the *variance* (σ^2). In any scientific experiment we try to get a reliable measure of the mean, based on a large number of measurements; and we hope that σ will be small, so that the mean is a really typical value. This is important not just to give us confidence about the accuracy of our

---

[63] 1 nanometre = 1 nm = 1/1000 000 millimetre = 1/1000 000 000 metre = 10^{-9} metre
[64] P. Hoffmann. 2012. *Life's Ratchet.* New York. Basic Books.
[65] 1 millisecond = 1 msec = 1/1000 of a second
[66] There are three kinds of average commonly used, the *mean*, the *mode,* and the *median.* What most people mean by the *average* value in everyday language is the *mean* and we shall talk only about the mean to keep things simple. To consider the other averages would not change the discussion significantly.
[67] Since some readers have asked, the way to calculate these statistics is shown in Appendices 1 and 2 at the end of the book.

measurement, but also because, if we want to compare measurements under different conditions, we need to use the variance to decide whether there is a real difference between the means. It is very important to note that without knowing the range of variability we cannot tell whether a difference between two means is real, or even how well a single mean represents data. For example, if all the measurements of temperature go up by 5 degrees in the Northern hemisphere and down by 5 degrees in the Southern hemisphere, the average taken over the whole planet is zero change! The way σ is calculated means that we can use it to express probabilities. It is often related to the shape of the *normal distribution* or *bell curve*. If a set of measurements is described by a normal distribution we expect to find about 68% of measurements within the range ± 1σ of the mean, and 95% of the measurements within ± 2σ of the mean. See Figure 5.1.

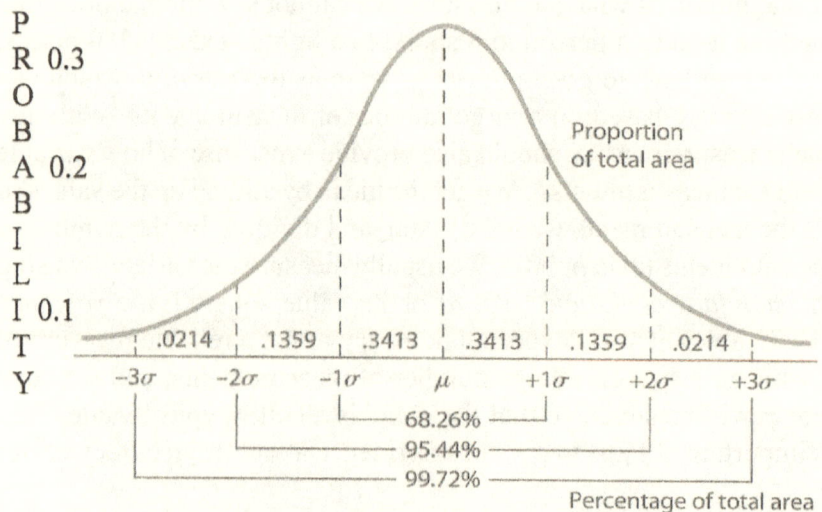

Figure 5.1. The normal distribution (bell curve).

Research on Drinking and Driving: a fable.

To see how probability relates to measurements of human behaviour, let's consider another fable, this time about an experiment on human reaction time. We won't go into all the details, either of how to do the experiment or how to do the calculations. But let's see in broad

Science, Cells And Souls

outline how probability comes into a scientific claim about behaviour. We ask volunteers to sit in a simulated automobile and drive through a simulated city. We take many measurements of the time it takes a driver to respond to a red stop-light appearing, and we find that the mean is 0.5 seconds. We then give the driver an alcoholic drink, take measurements "under the influence of alcohol", and the find that the mean is now 0.7 seconds. Have we shown that alcohol slows the response time to the red light?

Despite what looks like an increase, we can't decide just by looking at the means: the difference may have occurred by chance. Suppose we repeat the test 10 times, and look at the variability in these responses. To represent this variability we calculate the standard deviation of the measures. Suppose that the variability in the response times under the "sober" condition gave us a $\sigma_{sober} = 0.25$ seconds. From what we said about the bell curve on the previous page that means that even when sober there is a probability of $p = 0.17$ that the driver with a mean response time of 0.5 seconds will have a response time at least as long as 0.75 seconds on any one trial. So there is a probability of almost 1 in 5 of seeing a "sober response time" that is as long as the mean time under the influence of alcohol. Would we conclude from these results that alcohol slows the response time? Probably not. On the other hand, the mean$_{alcohol}$, the average of a lot of data, is certainly bigger than the mean$_{sober}$, and that makes us suspicious. If we found that the $\sigma_{alcohol} = 0.2$ seconds, then we could say that there is only a probability $p - 0.17$ of having a response time as short as the mean of the sober driver. So it seems somewhat "unlikely" (improbable) that in a sober driver we will see response times as long as the average times of the driver with alcohol; and it seems unlikely (improbable) that we will see response times from the driver with alcohol as short as the mean of the sober driver. What we want to know, clearly, is whether the differences between the means could occur "by chance". That is, supposing that we had tested the driver on two occasions when he was sober, could we expect to find a difference as big as

$$(\text{mean}_1 - \text{mean}_2) = 0.7 - 0.5 = 0.2 \text{ seconds}$$

by chance; just because "the dice came up that way"?

If we know how many trials (data points) make up each mean, we can do further calculations[68] and come up with a single number that tells us the probability that the pattern of data occurred by chance. Suppose we took 10 measures under each condition. Then taking into account the means, the σs, and the number of measurements, it turns out in this case that the result can occur by chance with a probability greater than $p = 0.06$. That is, there is a chance of about 1 in 16 of getting this result even if the alcohol had made no difference. So many people would probably say that while we may be surprised, it looks as though alcohol in the quantity we gave the driver did not have a statistically significantly effect on his behaviour.

Our conclusion is a statement about probability. The results might have shown that that there was a probability of 1 in 10, or only 1 in 20, or only 1 in 50 ($p = 0.1$, $p = 0.05$, $p = 0.02$) that the results were due to chance. At some point you would think, "Wait a minute. That is too low a probability for me to believe it is just by chance. I think the result is really due to the alcohol." At that point, although the result is still only a probability, you would regard it as a proven fact. *But the level at which you think a probability is a fact is up to you to choose*. That is true of *all* scientific investigations. The *p* level that summarises the statistics is called the *significance level* of the results. The smaller the value of *p* the greater, we say, is the statistical significance of the result. (Note that a high level of *significance* does not necessarily mean that the result is important in a social or intellectual sense, merely that it is very improbable that it occurred by chance.)

How do we choose a level of chance to treat as indicating that we have discovered a fact? The more important the outcome is, the more demanding the criterion we should use. Thus for a trivial question we might decide to accept that the result is true if the probability of getting the result by chance is less than 1 in 20 ($p < 0.05$), while for something that is a matter of important policy, such as deciding on a law about drinking and driving, we might want to use a value of 1 in 100 or even 1 in 1000 ($p < 0.01$ or $p < 0.001$).

---

[68] We could use the statistical test called the *t-test*.

Science, Cells And Souls

In theory, as we saw in an earlier chapter, we can never prove something to be true, but only prove it false. That is a fundamental matter of the philosophy of science and the logic of decision making. We can generally increase the significance of a result by collecting more data, and in practice everyone really does believe in the results of his or her research in a positive sense. I believe that in my research on attention I did prove that certain things are true about the nature of human attention. This belief is strengthened if other people repeat the experiment independently with similar results. If you are scrupulous you can think of the positive belief as a willingness to bet. If I say that my research is "true" at better than $p = 0.05$, I would certainly bet you £20 that I could get the same results again. If I choose $p = 0.01$, I would be prepared to bet you £100 that I could get the same results again, and so on. Much psychological research such as we will examine when we talk about neuroscience uses $p = 0.05$, and never would anyone accept a result as being secure if the probability that it is due to chance were greater than that. In fundamental experimental physics, such as work carried out at CERN, physicists often insist on significance levels of "5 sigma" because there is very little variability in the things being measured, they can collect very large amounts of data, and their measurements are extremely accurate. Behavioural and social scientists take significance levels of only "2 sigma" to count as facts, because their measurements are more variable and it is often impractical to collect enough data to reduce the p level. What is most important is repeatability.

Another question is where we get an estimate of the chance levels. In the fable of the research into alcohol and driving we needed the probability of observing a value of response time greater than 0.7 seconds, and we obtained that probability by empirical measurement. We calculated the σ of the $\text{mean}_{\text{sober}}$ response time from the data we had collected, and interpreted it as telling us about the expected probability of different response times. We compared that distribution with the one obtained under alcohol. So both the probabilities we were interested in came from empirical studies of the state of the world. We repeated the test so often that our estimates of the probabilities of different outcomes became sufficiently reliable. There is, however, another way getting probabilities against which to measure what actually happens, and that is to use a theoretical model. For example, we define the properties of a fair die as meaning that when it is tossed there is an equal probability of any one face coming up, and because there are six faces on a die and the die is defined as fair, the probability of any particular face is 1/6 (approximately $p = 0.1667$). We know this by definition: it is what we mean by a die being fair, that is having equiprobable outcomes for all its faces. We do not need any empirical tests. We can use this set of probabilities to see whether a die is in fact fair, or biased. We throw the die many times, measure how often each face comes up, and do the appropriate statistical tests to see whether these probabilities differ significantly from 1/6 for every face of the die.

We won't go into the mathematics here, but it is important that we can measure how real events differ from a theoretical set of probabilities in order to tell how the world works. We test against a theoretical model.

Since the expected probability of an event plays so important a role in assessing scientific data, it is worth noting two more characteristics of probability distributions. A bell curve represents data if the measured variable can take any value over the range investigated; that is if the measured property is *continuous* as mathematicians say. We have talked about experiments, such as throwing dice or using Zener cards, where the probabilities of events were all equal. If we draw a graph of the probability distribution of the outcome of throwing a die we do not get a bell curve, but a discrete distribution. See Figure 5.2.

A bell curve, or "normal" curve, will however also appear if there is not just one cause contributing to the occurrence of the event, but a large number of causes, independent of each other, even if each cause on its own has a discrete distribution.

Science, Cells And Souls

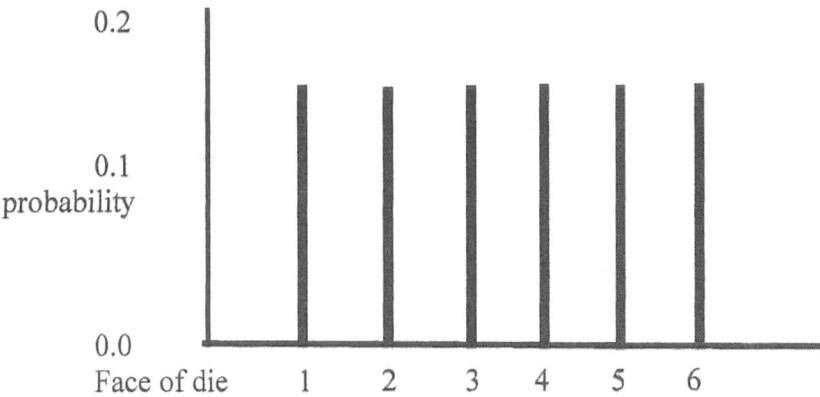

Figure 5.2. A discrete probability distribution. The horizontal axis shows the number on the face of the die. The vertical axis is the probability of each event. For a fair die each face appears with a probability of 0.167. There is zero probability of any number except those painted on the faces of the die.

Indeed, if you work out the frequency of different scores you can obtain if you throw one, two, three, or more dice at once, you will find that although the distribution is discrete and equal for a single die, the number of possible scores becomes more and more like a continuous bell curve as you add more dice. This is typical of biological systems. For example, the response time of a driver to a traffic light changing to red depends on where the driver is looking, how alert he is, how far his foot is from the brake pedal when the light changes, bloodflow through his muscles, and so on, including even the physical design of the vehicle he is driving.

As an example of judging against a theoretical probability, let's think again about experiments to see whether telepathy is a reality. As we saw, one way this research has been done is using a set of Zener cards, a set of cards each with one of five symbols on it. If the pack is fair after the shuffling, then the sequence of cards will have two important characteristics. First, the probability of *any particular* symbol will be 1/5 on each trial. Second, the *sequence* of symbols will be random; that is, the fact that one symbol turns up on a trial will have no influence on what symbol turns up on the next trial. If telepathy does exist, then we would expect to find certain characteristics in the data, the most important being that the symbol called by R will be closely related to, highly correlated with, the symbol sent by S. The probability that R says, for example, "circle" when S has thought "circle" will be greater than chance. If on the other hand telepathy does not exist, the situation will be just as though S dealt from a randomized pack of cards, and R dealt from another, different, equally randomized pack of cards. (It will be as though each rolled a fair, five-sided die independently of the other.) We can then measure the existence of telepathy by comparing the probabilities of the joint events we observe when one person mentally tries to transmit information to the other with what would happen if two random processes occurred jointly. Technically, we calculate the *correlation coefficient* or an equivalent statistical measure between the input and the output.

It is experiments like this that have repeatedly failed to show any statistically significant evidence for a relation between what is sent and what is received, so we have to conclude that telepathy does not occur between humans. Many different statistical tests have been used. Occasionally a study shows a very improbable run of successes, but any greater than chance results have proved to be unrepeatable. So we conclude that there is no real effect. This is a good example of how we can use a comparison between events in the world and a theoretical model of probability to help us decide what to believe.

Bayesian Probability

There is another approach to probability called *Bayesian*[69]. This does not talk about the frequency of events, but of the likelihood that a hypothesis is true given a particular piece of evidence. We start with an estimate of the probability that some hypothesis is true. We then get some evidence (data) about the hypothesis by observing the world, and estimate the probability of seeing that data if the hypothesis is, in fact, true. We combine these probabilities to get a final probability

---

[69] Named after the Rev, Thomas Bayes, 1701-1761, an English mathematician and clergyman, who is buried in Bunhill Fields, London.

that the hypothesis is true given both the initial probability and the probability, and value, of the evidence[70].

We saw earlier that we need to have better (stronger) evidence for things that are difficult to believe in the first place, and this effectively commits us to a Bayesian probability approach to the world. For example, I would demand stronger evidence to support the hypothesis that there is a ghost in my house than that there is a mouse. A Bayesian approach handles such situations and if we think in a Bayesian way we can better understand what we mean by scientific "truth". We perform an experiment several times and estimate the probability that the results may be due to chance, and hence the inverse probability that they have a causal explanation in terms of the hypothesis being proposed. We thus have a value for P(H), the probability that the hypothesis is true in the first place. Thereafter further experiments provide data from which P(H|D), the probability of the hypothesis given the data, can be estimated. This will increase or decrease depending on further results, and so our estimate of the probability of the hypothesis being true or false will progressively change, and may eventually approach 1.0 or 0.0 so closely that we can commit ourselves to the truth or falsity of the belief. It is in this sense that scientific method leads us to truth, and the discovery of new knowledge. Although we can never reach a probability of $p = 1.0$, absolute certainty, for our results, a Bayesian approach can lead us to a point where it is so exceedingly improbable that the results will be falsified by future repetitions of the experiment that we will accept the results as certain.

As an example, imagine tossing a coin repeatedly. Over the first fifty tosses we find that the probability of heads is $p = 0.53$. That is not unexpected even if the coin is completely fair. But if, over the next two or three hundred tosses the probability of a head does not come down towards $p = 0.5$, but climbs towards, say, $p = 0.55$ and then fluctuates about that value instead if about $p = 0.5$, the cumulative

---

[70] Equations for calculating Bayesian probability are given in Appendix 2 of this book for those interested.

evidence will make us eventually change from a hypothesis that the coin is fair to the certainty that it is slightly biased.

Some Properties of Probable Events.

The way in which probability impinges on our understanding of the world, and on the way we make decisions, is not straightforward, and humans are not good at dealing with probabilities. A common mistake is to think that if the probability of an event is low, it will not happen until a long time has passed. Thus, we hear statements such as the probability of a particularly high tide, or a particularly heavy rainfall, is such that it will only happen once in a hundred years. This is meant to comfort us, but should it? If an event happens by chance only with a frequency, on the average, of once in a hundred years we would certainly not expect to see many of them in a lifetime. But that does not mean such an event may not occur tomorrow. After all, if we toss a coin, it has to come up either heads or tails, and which one occurs on the first trial is not terribly surprising. If we throw a die, one of the six faces must come up, and the fact that the one on which we have bet comes up on the first throw is again not terribly surprising, even if in the long run it will only come up with a frequency of 1/6. Many people would be very surprised if they bet on the number 23 at roulette and it came up on the first throw; but after all, there is a chance of 1/36 that such will happen, and 1/36, or 0.028, is not a tremendously small number. On the other hand, we would indeed be very surprised if the number we bet on came up several times in a row[71]. It is not often we meet "the man who broke the bank at Monte Carlo"! We should always remember, when people say that an event is so rare that there cannot have been enough time for it to happen, that "improbable" means "infrequent", and does not usually say anything about how soon it will happen for the first time[72].

---

[71] For a compelling fictional account of such an occasion see D. Yates, 1922. *Berry and Co.* Ward Locke, London.

[72] There are certain kinds of probabilistic sequences in which rare events do not occur until a long time has passed, but these are special cases, and most events described as "one in a hundred years" or something similar are not among them.

That last comment is very important, because people sometimes say that life cannot have arisen by chance because there has not been time for the chance events to bring together randomly the components of life, and then for evolution to produce the higher organisms. Certainly, if the origin of life was a chance event we would not expect it to have happened often. But once only? And how rapidly do the events occur? There is usually no way of knowing for certain how soon something rare will occur for the first time just because it is improbable. For an account of the interaction of chance and necessity in the origin of life, see Hoffmann[73].

Uncertainty, and hence probability, enters the scientific picture of the world in many ways, but three are the most important for our discussion. First, at the deepest levels of physics, it seems that the nature of the material world is probabilistic. This claim is related to the discovery by Heisenberg of fundamental uncertainty in quantum physics. There is a limit on how exact our knowledge of the state of a particle can be. The more exactly we measure the momentum, the less accurate is our knowledge of the location of the particle; and the more exactly we measure the position of the particle, the less accurate is our knowledge of its momentum. There is a very small number, Planck's constant[74], that limits our simultaneous knowledge of the joint values of the momentum and position of a particle. There are several other pairs of properties that are also linked in this way, for example the duration for which something happens and the energy involved in the event. These properties of the material universe mean that for events at the level of atoms and smaller, we can only make statements of the probability that a particle will have a particular position and a particular momentum. This is not a limit caused by our ignorance of the true structure of matter: on the contrary, it is the result of our extremely *accurate* picture of reality.

Although Einstein hated the conclusion, and argued against it throughout his career, as far as we know there is no description

---

[73] P. Hoffmann. 2012. *Life's Ratchet*. Basic Books.
[74] One of the numbers fundamental to the physical description of the universe: $h = $ Planck's constant $= 6.626068 \times 10^{-34}$ m² kg / s

below quantum theory that can give a deterministic explanation of the uncertainty relations. The deepest understanding we can have about the physical universe is a set of statements associated with probable events, not certain events. The time-energy uncertainty relation means that energy (and hence indirectly matter) can appear out of nothing providing it only then exists for a very short time. So in some sense a chance origin of matter and hence of the universe may be plausible.

Let's leave aside quantum uncertainty and work only with very large entities like people and cars, where quantum uncertainty does not arise. We saw a second way in which uncertainty enters scientific investigation in our fable about an experiment on driving under the effect of alcohol. There is still uncertainty because there is variability in the measurements, and this must be represented as probabilities associated with statistical methods. Sometimes the magnitude of the variability can be changed by changing the kind of measurement that we make. For example, if we again think of Saltarella, the amount of variability, and how much it affects our conclusions about what event has occurred will change depending on whether we measure nerve impulses, or the position of joints or muscles, or the position above the ground of Saltarella's leg. Furthermore, if we limit our measures to the nearest 10 cm. there may be almost no variability, while if we measure in millimetres the variability relative to our measure will be larger. (One of the skills in doing research is to find an appropriate measure and to choose an appropriate accuracy.)

Finally, as again we have seen, uncertainty may be introduced into philosophical discussions as a theoretical construct, as in the description of the structure of events to be expected from the mathematics of a random sequence.

We have been taught that science gives us a true picture of the universe. How can that be reconciled with the probabilistic nature of fundamental physics? Well, one way is to go back to the notion of multiple Stories. Quantum physics is a story about the universe at very tiny scales, whereas other physics Stories are complementary to quantum physics and tell equally true stories at different measurement scales. So often science can tell us for certain why something happens. Planets orbit the sun because of gravity, and the way that gravity makes them move in their orbits is exactly described by Newton's Inverse Square Law, and Einstein's General Theory of Relativity. As long as we talk about things the size of planets on the scale of the solar system, and if the planets are moving far more slowly than the speed of light, even Newtonian predictions are pretty well exact. The predictions are deterministic. (Even when there are slight discrepancies these can be accounted for deterministically by Einstein's Theory of Relativity.) But sometimes scientific predictions may be only probabilistic. In general much of sub-atomic physics is probabilistic, and so are many more usual events, including weather forecasts, the economy, and the outcome of coin tosses or a roulette wheel.

It is important to understand how uncertainty, probability, cause, and predictability are related, because in the past some people have tried to find room for the concept of free will in the fact that scientific predictions are only probable. For example the neurophysiologist Eccles[75] suggested that acts of will may be able to affect behaviour due to Heisenberg's Uncertainty Principle, since physical determinism cannot apply to the atomic events in the nervous system, due to quantum uncertainty. We will look closely at the notion of free will in a later chapter and see that Eccles's idea is fundamentally wrong. For now let's concentrate on the relation of probability and predictability. The fact that science is probabilistic does not mean that prediction is impossible although it may constrain the way in which prediction can be done. We can predict that a coin will come up either heads or tails when tossed, and that a die will show one of the numbers 1 through 6 when it is thrown, even if we cannot say which will come uppermost on a particular trial. We cannot predict which atoms in a sample of radium will be the ones that disintegrate in the next second, but we can predict with very great accuracy what proportion of the radium atoms will disintegrate in a given period. We cannot predict exactly how many smooth skinned peas will be produced from a plant in a genetics experiment, but we can predict the proportion. We cannot predict exactly what the response time for the next trial in the

---

[75] J. Eccles. 1953. *The Neurophysiology of Mind.* Oxford University Press.

driving simulator will be, but we can predict to whatever accuracy you care to choose what the probability of obtaining behaviour with that value will be.

So even when we cannot scientifically predict *exactly* what will happen, we may be able to use a scientific theory or the scientific method to predict *completely* what will happen, that is the set of possible outcomes and their relative probabilities. And the proof of the ability to predict despite uncertainty is, as we saw in an earlier section, is that a prediction made at one level of description implies logically events at other levels of description, and also implies our ability to make technological devices that work as predicted to very great accuracy.

If we think back to Saltarella, there are different scales of measurement which are appropriate for different kinds of causes, and the accuracy of our scientific accounts are quite adequate to describe and explain the events in which we are interested. When they fail, we can explain the nature of the failures. And if we cannot do that, science gives us a method to set about discovering why the prediction failed. At any rate, there is no better methodology for discovering new facts about the world than science. As we saw, a great strength of science is that it is self-correcting if predictions are not accurate.

Attitudes to Probability

The psychology of human attitudes to probability is interesting. People find it extraordinarily hard to accept that events are random rather than caused. If presented with a series of random numbers and asked to account for the form of the sequence humans will create very elaborate explanations rather than say that the events are random. People have an enormous psychological resistance to the idea that events in a human life, or even in the universe at large, have no material or efficient cause, but are due to chance. Humans seem to have a deep dislike of chance and randomness, and there are several properties of random events that people find hard to believe. Sometimes researchers planning an experiment use tables of random numbers that can be found in many statistics books, but these are not always truly random. Most people subjectively underestimate the number of long runs of identical numbers that will occur in a truly random sequence. They think, for example, that a sequence of 4 identical numbers in a sequence of numbers drawn randomly from 0 to 9 means that the sequence is not random. But in fact, if we print 80 truly random numbers to a line and 30 lines to a page, we would expect a run of four identical ones about once every two pages. Sometimes computer programs are used to generate random number sequences, but again, because computers are deterministic, these sequences are not truly random. The only way to obtain a truly random sequence is to use some physical process that is completely stochastic, such as the intervals in a sequence of decays of radioactive atoms. People think that if a lottery is fair, it is more likely that a number such as 496532 will win than a number like 111111: but if the lottery is truly fair there is no difference in the probability of the two numbers winning.

Onwards to Human Nature

These chapters may have seemed dry, and a long way from our promised exploration of human nature. But they are important. We must understand what science claims and how certain it is before we can see how to treat scientific discoveries. We are now in a much better position to tackle the relation between different kinds of knowledge. There may be several accounts of any event, equally true. The meaning of what we investigate can be examined analytically by philosophy and empirically by science. It is not always the latest knowledge that is most revealing. And science is based on probability but can tend to certainty. We now have the intellectual tools we need. Let's use them directly to examine what it is to be human. How can both science and philosophy be brought to bear on the Fundamental Words, on our understanding of human nature?

PART 3
THINKING ABOUT HUMANS

Chapter 6

Names, Nouns and Things

In principio erat verbum[76].

The Gospel of St. John

How often misused words generate misleading thoughts.

Herbert Spencer. *Principles of Ethics.* 1879.

We are going to see what science has to say about the *Fundamental Words,* about life, intelligence, the mind, and consciousness. But what exactly do those words mean? In recent years a lot of new scientific research has appeared, but we have to be careful. Think back to the Electrician and the Advertisement. In that fable a scientist, the Electrician, was able to analyse all the physical properties of the neon sign, but was unable with his scientific equipment to tell that it was an advertisement. A scientific account does not say everything there is to say about a physical object. We know that the Fundamental Words are rather special, and that people have struggled to understand their meaning for centuries. Could it be that some of them are more like advertisements than neon tubes? Is that why it is so hard to understand what science has to say about them? We don't want to waste our time by misapplying scientific methods to phenomena to which they do not logically apply; nor do we want to reject science because we misunderstand the nature of what we are investigating.

Language can be misleading. Perhaps some of the mysteries that confront us when we think about human nature are not, in a sense, real difficulties. Perhaps they don't even need science to solve them. If we look carefully at how we use language, the puzzles may

---

[76] In the beginning was the word.

disappear. So let's look at some linguistic traps that await us in the Fundamental Words. When we talk about human nature we think from personal experience that we need words like *life, soul, mind, will*: we think they refer to basic aspects of being human. But what does that mean? What do they really refer to? Are there really *parts* of humans of which they are the names? Is *life* or *mind* a property, a part of a human, a *thing* perhaps made of a mysterious non-physical kind of stuff? Or is there a better way to think about such words?

It is a characteristic of English that nouns are often the names of *things,* such as *dog, flower,* and *brain.* Since much of our world is full of things, we may come to think that all nouns refer to things. Then it is natural to ask, "How is the mind connected to the body?"; or, "Where in the brain is consciousness located?" It seems that these are questions about *things,* namely *minds, bodies, brains,* and *consciousness.* But do such *things* really exist?

The sentence, "The cat is sitting on the mat." probably makes us think that at least two things exist, namely a *cat* and a *mat.* (Note, by the way, that we don't feel the same about *the, is, on,* or even *sitting.*) Furthermore the sentence looks like the answer to the question, "Where is the cat?" For nouns that are the names of concrete objects that is a perfectly reasonable attitude and is not misleading. Take another example. To say, "The tap is connected to the nozzle by a hose.", makes one think of three things, a *tap,* a *nozzle,* and a *hose,* and looks like the answer to the question, "How is the tap connected to the nozzle?".

The rules of language state that strictly speaking *concrete* nouns are the names of *things,* for example *dog, cat, brain.* But there are many "things" that have names that are nouns but which do not exist concretely, for example *bravery, kindness,* and *generosity.* We don't expect to be able to walk down a street and say, "Oh look! There's a lump of bravery!", or "I wonder why they've left all that kindness lying about." What we might say, rather, is "Oh look! What a brave thing he did just then!" or "That was a kind act."

Even what look like relatively straightforward words can be difficult. Consider the names of colours. When we say that a lemon is yellow, what do we mean? Well, things are not yellow because they "have yellowness". Having ice on it may make a path slippery, but having yellowness is not what makes a lemon appear yellow: at least not having yellowness in the same sense as the path has ice. Although Plato seems to have thought that there is an essence of yellowness that exists somewhere and of which all yellow things partake, that is a mistake. When we say that something is yellow what we are really saying is that it is like other things that we also call yellow. We are drawing attention to the fact that it is appropriate to describe it as we describe other things that are like it in respect of its colour, and that we have learnt how to use the word *yellow* correctly in our language.

In the case of *yellow* we can make this clear by referring to a Scientific Story about colour. Research into colour perception has revealed that there are an infinite number of ways of making a patch of light on a white surface appear yellow to a viewer. Light can be thought of as a wave of electromagnetic energy, and what we experience as colour depends on the wavelength (or frequency) of the light. If we make light waves of only one wavelength we get a light that we experience as a pure colour. Light with a wavelength of about 400 nanometres (nm) appears to us as deep blue-violet, while light of wavelength about 700 nm appears as deep red. The yellow light from sodium vapour, the yellow that we are familiar with from street lighting, is an almost monochromatic light with a wavelength of 589 nm. Any white surface illuminated with sodium light at 589 nm will appear pure yellow. But the light does not contain "yellowness". Indeed, because of the way in which the eye and brain process colour information we can take a monochromatic red light of about 700 nm and a monochromatic green light of about 500 nm, *with no other wavelengths present,* and if we mix them at carefully chosen intensities what we will see is a yellow that is indistinguishable from the 589 nm sodium yellow. The two are indistinguishable for all purposes: if we mix the sodium light with, say, a blue light, and mix our red+green with the same blue, the mixtures will be indistinguishable. If we make each brighter by the same amount, the brighter lights are indistinguishable. The two yellows are absolutely indistinguishable: and yet one has no yellow light in it at all! So "being yellow" is not a matter of being given "yellowness" by something that contains it. It is more about learning how to use colour words correctly to categorise one's experiences.

Now while an empirical Scientific Story brings out the way in which *yellow* is not the name of a *thing* possessed by yellow things, I have used the Story just for emphasis. We could have conducted the discussion entirely as a Philosophical Story, analysing the way the word *yellow* and other colour words are used in our language. But the point remains: nouns are not always the names of particular things, and may indeed not be the name of anything. There may not exist any *thing* of which a noun is the name, even though the noun exists in our language, for example a basilisk, or a unicorn[77].

When nouns refer to abstract entities we can be misled. At first sight the question. "How is the mind connected to the body?" looks as though someone is talking about two things, a mind and a body, and is expecting an explanation in the form of a sentence about those *things*. The sentence would presumably be something like, "The mind is connected to the body by. . .". By what? Obviously not by a hose! Not even, as Descartes thought, by the pineal gland. What *sort* of answer could there be? What *kind* of Story? Or try asking, "Where is consciousness?" Obviously not on the mat! Perhaps nowhere, even though we are usually conscious. Again, "How is my will connected to my muscles so that I can voluntarily act?" What could possibly be an answer? If it is so difficult to see where to look for an answer, perhaps there is no answer. Perhaps the mind and the will are not things at all. But if that is so, why do they play such a Fundamental role in describing what it is to be human?

There are many sentences that have the grammatical form of questions, but are not really questions. . . and *therefore* don't have answers! There is a famous sentence in modern linguistics that states, "Colourless green ideas sleep furiously." Let's turn it into a question. "Do colourless green ideas sleep furiously?" The answer is neither "no" nor "yes", because the sentence is not really a question. True, it looks like a question, and ends with a question mark. But the phrase "colourless green ideas" is strictly meaningless, since colourless things cannot (logically) be green, and ideas cannot be (logically)

---

[77] The beautiful long ivory horns that are sometimes found are of course the tusks of narwhals, not horns of a traditional unicorn horse.

Science, Cells And Souls

coloured. Nor are ideas the kind of things that can be asleep or awake. So what looks like a question actually is not. *That* is why it is difficult to answer – because it is not really a question, not because the question is particularly difficult. Is the difficulty in talking about the Fundamental Words perhaps this kind of difficulty?

In order to understand human characteristics perhaps we need to change the way in which we pose the questions, so that we can be sure that they do, in fact, have answers. Consider the following two sentences:

1. "What is the mind and how is it connected to the body?"
2. "Under what conditions do we say that someone's bodily behaviour shows that mental events are taking place?"

Question (1) is typical of how people have often approached the so-called "mind-body problem", and no acceptable answer to it has been proposed. Question (2) means more or less the same thing and is more approachable. We will investigate it later in this book. The same approach, making the noun into an adjective, applies to the following pair of questions:

3. "How does my free will make me act?"
4. "Am I really responsible for my actions?"

In the case of both Question 1 and Question 3 what seems to be a question without a meaningful answer can be turned into one that at least looks promising as an approach to human nature by changing a noun into an adjectival phrase, and thus avoiding the temptation to think that nouns are the names of *things*. Remember – the advertisement in Chapter 3 was not a separate *thing* that was connected to the electrical sign, but an *ability* the sign possessed only if the culture recognised it. The role of the Fundamental Words in describing human nature is much clearer if we start from the assumption that like *yellow*, they are not the names of things possessed by humans, not parts of humans, but labels for ways of talking about what humans do, ways of describing the biography of humans. To adopt such an attitude will surely affect both how we do research and how we interpret it.

History, Language and Translation

Our ideas about human nature have many intellectual ancestors. These include Greek philosophers such as Plato and Aristotle, late Roman and mediaeval philosophers such as Plotinus, Augustine, and Aquinas, philosophers of the Enlightenment and the 18th century such as Descartes, Locke, and Hume, down to modern philosophers such as Russell, Wittgenstein, and our contemporaries. The Fundamental Words occur in many languages and have been translated again and again, in our time as *soul, life, mind, intelligence, consciousness* and *will*. But have they always had the same meanings? And has the way they relate to the rest of language stayed constant?

It is a mistake to think that modern discussions are inevitably more advanced than those of the ancients. To give but one example, Bertrand Russell once said that Peter Abelard[78] in the 12th century had discussed a major problem in the philosophy of language with as much insight as any modern philosopher. The philosophical discussions that have come down to us are often clever and subtle, even if the science of that time was relatively primitive. We can often learn a lot from them[79]. So let's take some time to see what our intellectual forbears said about human nature.

One problem for us is that most of the earlier discussions were in languages other than English. We use the word "mind" or "soul". Descartes wrote of *"l'ésprit"* or *"l'âme"* when he wrote in French, and *"anima"* when he wrote in Latin. Aquinas and others writing in the mediaeval period used *anima* to translate ψυχη (*psyche*) from Greek often through an intermediary translation from Greek to Arabic and then from Arabic to Latin. There was a Latin word *mens*, equivalent to the Greek νους (*nous*)[80], from which our word *mental* is

---

[78] He also anticipated Frege in developing formal truth table logic, and described a version of Pascal's Wager about ethics. See Peter Abelard in the *Online Standard Encyclopedia of Philosophy*.

[79] For example, as we shall see later, mediaeval philosophers thought it was fairly easy to create living from non-living matter. Their science was poor, but their philosophy was good.

[80] As in the phrase, "Use your *nous*".

derived. We need to look at the way translation may cause problems because translation is not the same as transcription with a dictionary. Even if the ideas were originally sensible and useful, perhaps their meaning has been degraded over the years as they have come down to us. To understand what others said we need to understand more than their language, for as Wittgenstein said, "To imagine a language is to imagine a way of life": colour names which in one culture may summon up thoughts of joy and liveliness in others are associated with death and mourning.

Consider the problem that philosophers call "the Cartesian *Ego*", that is the "*Ego*", the "*I*", that was described by Descartes. Descartes, looking for something he could not doubt, famously settled for *"Cogito, ergo sum."*, generally translated from the Latin as, *"I think, therefore I am."* In English we replace *Ego* by *"I"* or *"Me"*, and then follow Descartes's discussion in the form, "I can doubt that the world exists, or that my body exists, or even that my mind exists; but at least I cannot doubt that *I* exist as long as I can think. Because even if I doubt that I exist, that is a thought; and something must exist in order to doubt that it exists. So in the end I cannot doubt that I exist." Descartes then went on to ask, "What can I say about this "*I*" whose existence I cannot doubt?" He concluded that a human being is a "thinking essence".

For the moment I don't want to worry about whether Descartes's line of reasoning was valid or not: I just want to look at how language works in this kind of discussion. In English we use "I" to translate *"Je"* or *"Moi"* from French, or *"Ego"* from Latin. It seems quite reasonable to follow up the sentence "I think, therefore I am", by asking, "Then what is this *I*?"

Here is what Descartes said, writing originally in French:

> Après y avoir bien pensé, et avoir soigneusement examiné toutes choses, enfin il faut conclure, et tenir pour constant que cette proposition : "Je suis, j'existe", est nécessairement vraie, toutes les fois que je la prononce, ou que je la conçois en mon esprit. [...] Je ne suis donc, précisément parlant, qu'une chose qui pense [...] C'est-à-dire une chose qui

doute, qui conçoit, qui affirme, qui nie, qui veut, qui ne veut pas, qui imagine aussi et qui sent.⁸¹

But the Latin used by Descartes for "Je suis, j'existe" was, "*Cogito, ergo sum*", a sentence in which "*Ego*", the Latin equivalent of "*I*" (in English) or "Je" or "*Moi*" (in French), does not appear. That is because the way Latin verbs function is different from the way verbs work in English. Instead of using pronouns to distinguish who is acting, ("*I* think", "*You* think", "*They* think") in Latin we simply change the ending of the verb, ("Cogit*o*", "Cogit*as*", "Cogit*ant*")⁸². If you read a classical Latin writer such as Pliny, you can read many pages of his letters, written to his friends or to the emperor, without ever coming across the word "Ego". He seems to keep the latter for a special emphasis. He might say, "Ad urbem ambulavi." to mean, "I walked to the city"; but if he wanted to emphasise that it was indeed he, Pliny, who went to the city, he might say, "Ego ad urbem ambulavi", to mean, "I myself walked in person to the city." In Pliny's messages to his friend Cornelius Tacitus the word *"ego"* only occurs 14 times in 24 different letters, which total about 5700 words. Try writing 24 letters to a close friend in English while avoiding the word "I" and you will see how different are the languages and yet how you use them to say the same thing. Pliny uses *"ego"* almost always for emphasis: "Quis? Ego sed nihil refert." ("Who? Actually it was *I*, but it doesn't matter.") "...nam tu magister, ego contra; atque adeo tu in scholam revocas, ego adhuc Saturnalia extendo." ("..for *you* were the leader, not *I*; and while *you* went to school, *I* used to run off to the festival.")⁸³

If I think about my nature in a language that does not need to use the pronoun "*I*" to mark the fact that *I* am thinking, perhaps I would

---

[81] "After I have thought hard and carefully examined everything, in the end I have to conclude, and consider absolutely certain, "I think, I exist" is necessarily true whenever I say it or think of it in my mind...So strictly speaking, all I am is something that thinks...that is something that doubts, conceives ideas, affirms, denies, wishes, refuses to wish, and also that imagines and has sensations."
[82] In French we do both: "*Je* pens*e*", "*Tu* pens*es*", "*Ils* pens*ent*".
[83] Pliny, *Letters*. Loeb Edition. Harvard University Press.

not find it so natural to go on and ask the question, "But who is this "*I*", really, who does the thinking?" If I say in Latin, "Dico", a standard translation would be "I am speaking." Obviously it is not the immaterial, internal Cartesian *ego*, the "*I*", who is the subject of the verb, but I-the-person-as-a-whole. (If it were the immaterial *I*, no one could hear me because I would not have any physical throat, lungs or vocal cords with which to speak.)

As we have seen, in Latin you usually don't use the word "I". Curro. Festino. Dico. Puto[84]. These don't need "*Ego*", so one is not tempted to ask, "If there is an ego which features in every word, (ego curro, ego festino, ego dico, ego puto) then who or what is this ego that they all have in common?" In fact, if you did ask that question, the answer in Latin would be as in English – "It is in each case the same *person, the one who is the subject of the verb.*"

Consider the following.

1. "Who's running? Stop or I'll shoot!"
2. "It's I, Neville Moray not the murderer who's running: so don't shoot."

If I thought that "I" referred to my inner ghost, I wouldn't be worried because a bullet can't hurt a ghost, and equally there would be no point in your shooting. It is only if "I" refers to the whole person that the conversation makes sense.

3. "Who's thinking? Can I trust the answer?"
4. "It's I, Neville Moray, (not Descartes) who is thinking, so of course you can!"

Since language works in the same way for the examples (1) and (2) on the one hand and (3) and (4) on the other there is no reason to think that changing the verb from *run* to *think* has altered anything: after all, we often think aloud.

---

[84] I run, I hurry, I say, I think

Actually, even in English we do sometimes use a form like the Latin, without the personal pronoun. It is a dialect we can call *"Diarese"* because it is used traditionally in Diaries!

> "Got up early and had breakfast. Went to the Museum and looked at the Botticellis. Drank a coffee. Thought about my book. Decided *I* should write the preface, and not ask Tim Firth to do so. Thought about Descartes. Can imagine thinking without a body, but who would then be thinking? Must still be Neville."

In that passage of diarese English is very like Latin, even to the use of *I* to emphasise the writer as distinct from anyone else. But there is nothing to tempt one to ask, "Then who is this *I*?". In fact, it makes it very obvious that *I* am just the person sitting at the desk typing on my computer.

The way our language works can bias the way we think about things. So when we read translations of what earlier philosophers said about *soul*, or *life*, or *mind*, or *will*, that is (in Latin) *anima*, or *vita*, or *mens*, or *voluntas*, we need to know whether the translator was just looking up what was written in Latin in a dictionary, or whether he or she chose the English words carefully in the light of a deep understanding both of how the people of the time used the words, and also how language was used in the translator's own lifetime.

Here is a table to show how some of the Fundamental Words correspond to each other in different languages. The table is incomplete, but that is because words are often used loosely in ordinary language.

ENGLISH	**FRENCH**	**LATIN**	**GREEK**
the life	la vie	vita	ho bios 'ο βιος
the soul	l'âme	anima	he psyche 'η ψυχη
the mind	l'esprit	mens	he nous 'η νους

the will	la volonté	voluntas	no equivalent in ancient Greek
the man	l'homme	homo	ho anthropos 'ο 'ανθροπος
		vir	ho aneer 'ο 'ανηρ

We can see at once that there will be cases where it may be hard to do a good translation. For example, Latin has two words both of which are translated naturally by *man* in English: but *homo* usually means *mankind*, while *vir* means *man* as opposed to *woman* (*mulier*), although in Latin *homo* is sometimes used just for *man*. Greek has the same distinction between *anthropos* and *aneer*. Another difference is that Latin does not have definite and indefinite articles, so (to choose an example that was much in the news a few years ago,) Prince Charles can legitimately translate the British monarch's title *Fidei Defensor* either as *Defender of Faith* or *Defender of the Faith*, unless he takes account of the context in which the title was originally awarded by the Pope to Henry VIII, in which case only the second is correct.

Another problem arises because the meanings of words shift as time passes. An obvious example in our time is the word "gay" in English. In the 1940s and the 1950s the only meaning of this word was as a synonym for *happy, joyous, full of fun*. At present the only meaning in Everyday English is *homosexual*. Another classical example is *nice*. Most people know that in the 19th century the meaning of *nice* was *precise*, as in the phrase *a nice distinction*. The Oxford English Dictionary tells us that several centuries earlier *nice* meant *lascivious*[85]. We inherit Stories that were often written when words had different meanings; and it is important to understand the meanings words had at the time the Stories were first written if they are to help us to understand human nature.

---

[85] Which for an etymologist can be a source of quiet humour when a friend says, "I am so glad my son married such a nice girl."

Where does this leave us? Looking at language as we have done in this chapter leaves us with a new starting point for examining the Fundamental Words. On the one hand, there is no doubt of the power of scientific, physicalistic, material accounts of human nature wherever they are appropriate. But part of human nature is the way in which language has evolved and provided us with ways to think about problems. There are serious difficulties in reducing a description of humans to a simplistic materialism, but equally with the suggestion that immaterial parts of a human exist, although some human characteristics are not "just" material. Are perhaps the Fundamental Words related to scientific language as the advertisement was related to the Electrician's language in Chapter 3?

To explore human nature means that we have to keep a balance between the methods of science and the methods of philosophy. Indeed we should look at the ways in which all the kinds of Stories complement each other. Each has a role to play in understanding human nature, and omitting any of them means the overall description will be deficient in some way. In order to understand what science has to say about the Fundamental Words we must be careful to make clear to what kind of entity the Fundamental Words refer.

So let's look at what Science has to say about the Fundamental Words.

Chapter 7

Life

I've looked at life from both sides now.

Joni Mitchell. 1967.

The Science of Life

Humans are *alive*. So what is *life*? How did it arise? Can we create something that is alive? Remember that as we saw in Chapter 6 a word like *life* may not be the name of a *thing*, something that is possessed by living things but that non-living things don't have. Certainly being alive is rare and special. There are very many things in the universe that are not alive, including rocks, water, stars, and (curiously!) even the chemicals, atoms and sub-atomic particles that make up the bodies of living things. We even naturally distinguish between non-living things and dead things. Only things that have once been alive can be dead: other things are just non-living. And when people talk about computer programs imitating life that is just a metaphor, used to describe a program that makes little icons reproduce on the screen, and is nothing directly to do with real life. So far all attempts to write programs that spontaneously reproduce and evolve have failed[86].

We sometimes say that living things *are alive*, and sometimes that they *have life*. The first of these phrases is harmless, and emphasizes that to say something is alive is a way of drawing attention to its biography, to what it does, how it behaves. But the second phrase is dangerous. It may make us think that there is some *thing*, or *something*, that a living body *has* that a dead one or a non-living

---

[86] C. Venter, *Life at the speed of light*. 2011. London. Little, Brown.

thing does not have. We would have to inject such a thing into non-living matter to make it live. Some people say that what a human body has when it is alive is a *soul* which goes away, free from the body at last, at death. Others said that *all* living things have souls – rhubarb, aubergines, hippos, and humans, but that their souls are of different kinds. We'll look at that idea later on. Descartes thought that animals other than humans were just elaborate machines, and that their suffering, pleasure, etc., were mere illusions. Aristotle thought they all had souls of different kinds. If making something alive really involves putting a soul into it we may be going to have a difficult time. But is this how to look at the problem?

Let's start again. Why should we think that something leaves a body when the latter dies? If we want to decide whether something is alive, we don't go looking for the presence of some mysterious substance, *life*, or for a soul, that guarantees that a creature is alive. Rather we do exactly the opposite. We look at what a creature looks like, how it behaves, and how closely its biography resembles that of other creatures that we think of as alive. If it looks like a living creature and behaves like a living creature, then we say, "It's alive"[87]. We may sometimes vary the way we talk for convenience, or for aesthetic reasons, and say that *it has life*. But to do so does not add anything to the first phrase. It says the same thing in a different way. We need to reject the idea that life is a substance, some *thing* that a living creature *possesses*. The only way we know that something *has life* is to decide that it *is alive*. It is the adjective, not the noun that is important. Similarly, the only way that we can decide that something has died is because it stops behaving in the way that living things behave. We never see *life* leave it. We just decide, on the basis of what we can observe, that words to describe the actions and properties of a living thing no longer apply. When we decide that something has died, (even a human being,) we make a decision about what kind of language is appropriate to describe its state and its behaviour. It is not a matter of seeing something depart.

---

[87] "If it looks like a duck, and walks like a duck, and quacks like a duck, then it's a duck."

To decide whether something is alive or dead can be difficult, and in some cases we may be uncertain. That is why ethical problems arise when people are in a so-called "vegetative state" after an accident or a stroke. Some of the signs of living are present, but not all; perhaps not very many. How many properties, and which, do we feel are enough and of the right kind to let us conclude that a person is still alive? And since deciding that something is alive is a matter of looking at how it is behaving, sometimes it is very difficult to decide whether non-human things are alive. Many of them don't really interact with us in obvious ways. "Stone plants" that live in deserts do so little, so little typical even of plant-like things, that they hardly seem at first sight to be alive at all, although a close scientific examination shows that they have the structure and metabolism of plants, and therefore are alive. Seeds, tardigrades, and the spores of bacteria may seem dead or non-living for long periods until a change in their environment allows them to show behavior such as metabolism and reproduction that we know implies that they are alive[88]. And some things, such as viruses, can even be crystallised, and are so borderline that no one has really come to a decision one way or the other – not because we cannot tell whether or not they "contain life" but because their biography has little in common with other living things and rather more with what we think of just as complicated chemicals.

Imagine visiting another planet and discovering things that move, take in energy, reproduce, and so on, but bear little or no resemblance to creatures on earth. Perhaps they are not even made of carbon. Would we say they are alive? Think of the specimen that some researchers said is evidence of past life on Mars. It was found in the Antarctic in a meteorite that came from Mars. Some researchers described it as a "fossil bacterium" because it looks like fossil terrestrial bacteria: but other researchers believe it is just a peculiar form of rock. Another theory is that it is an artifact of how the samples were prepared for examination under the microscope, and overall there is not enough evidence, one way or the other, to decide. The only way we can decide

---

[88] The island of Gruinard in the Hebrides was inoculated with anthrax in the 1930s and the spores in the soil were still considered dangerous at the end of the 20th century.

whether things are alive is by what they are made of, what they look like, and what they do. We cannot find life in them and *therefore* conclude that they are alive. You may know a children's story called *The Thirteen Clocks*. In it there is something called the Todle, which a character describes as "a blob of glup" that makes sounds like kittens screaming. He tells another character that if it catches him, "The blob will glup you."[89] Is that enough to make us say that the Todle is alive?

Almost certainly scientists will soon claim to have made a living system in a laboratory. Already forms of DNA that had not previously existed have been synthesised and injected into cells from which the nucleus has been removed, and the resulting organism has reproduced[90]. Judging by the tone in which people discuss problems such as stem cell research, cloning, the nature of DNA and other complex biological processes in the media, the claim that life has been created in a laboratory will be viewed with horror by many people. But why should that be so? What would the claim mean? Is there a fear of violating God's unique right to make life, or is the fear a rational worry about the possibility of releasing uncontrollable pathogenic organisms into the environment?

If we try to synthesize life that means that we are going to try to combine chemicals to make compounds that behave in certain ways, entities that reproduce themselves, draw their energy from the environment, and (to talk technically for a moment) show an ability to resist entropic degradation[91], death caused by the dispersal of useful energy and the loss of organization. The feeling that scientists should not meddle with life is not confined to religious people: it seems widespread in our society. Often people talk of the danger of creating a "Frankenstein's monster" with terrible consequences. (Few people seem to have read *Frankenstein*[92] and so realize that the story is not

---

[89] J. Thurber. 1992. *The Thirteen Clocks*. Yearling. I need hardly say that with language like this children love the story.
[90] http://www.wired.com/wiredscience/2010/05/scientists-create-first-self-replicating-synthetic-life/. C.Venter, *Life at the speed of light*. 2011
[91] We will discuss the meaning of this in shortly.
[92] M. Wollstonecraft. 2003. *Frankenstein*. London, Penguin Books.

a horror story so much as a tragedy about the fate of the unfortunate monster.) There seems to be at one and the same time a refusal to believe that it is possible to create life, and also a great fear that it will happen. Many religious people seem to feel particularly threatened by the idea, even though there is nothing in the Christian religious revelation to suggest either that life cannot be synthesized or that we should not try to synthesise it[93].

Many people seem to feel that life is so special that it must be in some sense supernatural. But at the same time people who feel reverence for life usually restrict their reverence to certain kinds of living things: plague-bearing rats, malaria-bearing mosquitos, pathogenic bacteria, the HIV virus, ebola and other vectors of disease don't get much consideration from people who revere "life" as such. And while many people talk of life as "sacred", or so special that we should not interfere with it in "unnatural" ways, they almost always seem to be thinking of life as *something*, a substance, which certain entities *have*. Such attitudes are strange, because in contrast to the widespread modern belief that to synthesize life is either impossible or undesirable, for most of history people thought that to create life is relatively easy, happens quite frequently, and is a goal worth aiming for. People have doubted this only for about the last 150 years. The scientific evidence that people in the past adduced for their belief that it could be done was completely inadequate, although it satisfied people at the time. But they did not see a philosophical difficulty about making life from non-living matter.

For many centuries magicians and alchemists tried not just to make living creatures from non-living materials, but even to make humans. It was taken for granted that if you left dead material in a suitable environment new living creatures would arise from it: maggots could be created from dead meat. The Latin poet Virgil tells us that swarms of bees arise from dead animals, and in the Bible we read that "out of the eater came forth meat, and out of the strong came forth

---

[93] Even if the Bible is read literally to say that God created living things, it does not say that man cannot do so. (For example, making life is not one of the things that God claims Job could not do.)

sweetness"[94]. A report to this effect can even be found in an early paper to the Royal Society in the 17th Century, and it was commonly believed in the Middle Ages. So why is it now Everyday Knowledge that living things cannot be made from non-living components?

The answer lies in a 19th century scientific dispute involving Louis Pasteur. While doing the research that led to the development of "pasteurization" he became embroiled in a controversy about whether living cells were always necessary for fermentation and putrefaction. His experiments seemed to prove conclusively that they were; and because the controversy had come to involve also a controversy between religious and atheist scientists, the results unfortunately were taken to mean that living cells could not be created from non-living materials. What the research showed, of course, was that living cells of the kind that Pasteur had investigated could not be created from the kind of sterile nutrient media he used just by leaving the latter lying about for a long time. But that is quite a different thing from saying that no living organism can ever be created from non-living materials. And subsequently fermentation has been induced without the presence of living cells. It is somewhat ironic that in the Middle Ages, often called the Ages of Faith, people thought you could create living from nonliving materials, on the basis of poor science but good philosophy; while in the 19th century when religion was under attack they decided that you could not create living from nonliving materials on the basis of good science but poor philosophy. After all, we turn nonliving matter into living matter every day when we eat sugar, and dead matter into living matter when we eat kippers or sausages!

Be that as it may, let's think about what is involved in making something alive. First, remember that we are not trying to insert a quantity of *life* or an immaterial soul into some inert material, but asking how we might manipulate nonliving materials into a form such that when we look at the biography of what we make we would

---

[94] This quotation, with an illustration, can be seen on any tin of Tate and Lyle golden syrup.

Science, Cells And Souls

decide, "Yes, this is alive. It metabolises. It reproduces. It is sensitive to the properties of the environment."

What is it that makes us think something is alive? Whatever it is must apply equally to amoebae, elephants and people. If we look at large organisms such as people, we are certainly not looking at the simplest kind of living things. If we examine a living body, we don't just find a random collection of molecules. The human body is made up of *organs* such as skin, blood, kidneys, heart, brain, and pancreas. Each of these in turn is composed of smaller components called *cells*, and the structure and function of cells differ depending on which organ we examine. A cell is a very complex structure, bounded by a membrane, a kind of fatty sac that prevents the contents from spilling out, and prevents materials from entering the cell unless they have certain physical properties. Cells function as biochemical factories using organelles such as ribosomes, mitochondria etc., within which chemical reactions take place. To see details of the complexity of these operations, and the amazing ways in which they are embodied in autonomous biochemical "robots", see the wonderful book by Hoffmann[95]. Growth and reproduction occur when cells divide, having derived the necessary energy from chemicals and other sources of energy in the environment. (The environment of a cell deep in the human body is, of course, the other cells and body fluids that surround it. In a plant the environment includes sunlight.) For a schematic version of a cell see Figure 7.1.

A human is made up of hundreds of different kinds of cells, each specialized to perform some function to support the whole organism's staying alive. Some cells are constantly being renewed: the cells that make up the skin, for example, do not live for more than a few weeks. Others, such as the ova in the female from which a new organism can develop, are all present soon after birth, and nerve cells are not normally replaced if they die or are severely damaged. The structure and function of cells differ greatly from organ to organ (a nerve cell could never be mistaken for a liver cell for example,) but all cells in large animals are variants of two basic types, called technically

---

[95] P. Hoffmann. 2013. *Life's Ratchet.* New York: Basic Books.

eukaryotes or *prokaryotes*, depending on whether they do or do not have a nucleus. In the human body the vast majority of cells have nuclei: almost the only ones that do not are red blood cells.

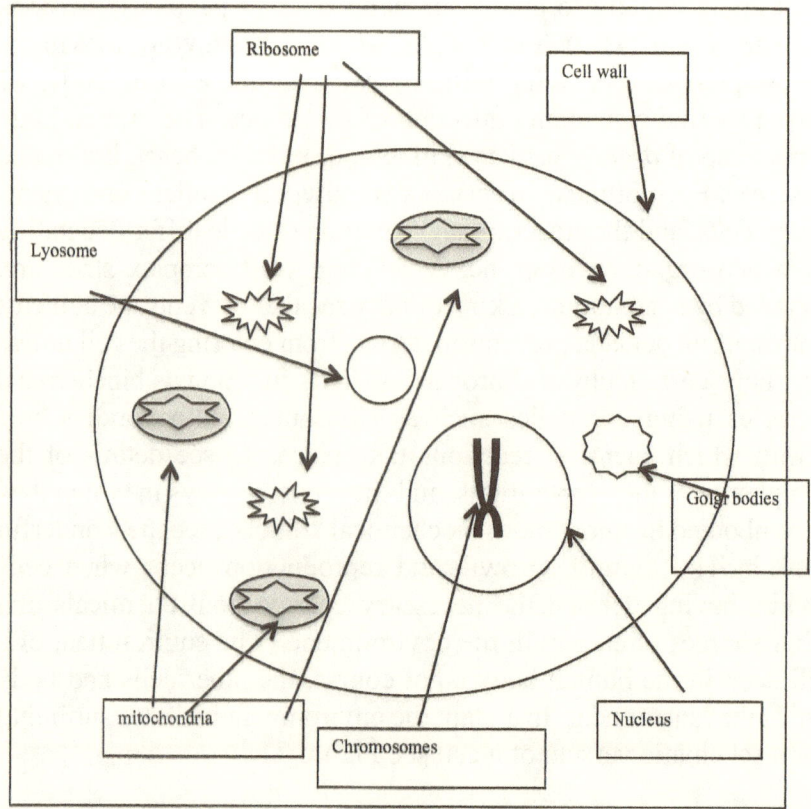

Figure. 7.1 Schematic representation of an Eukaryotic cell. Typically a cell is about 0.01 to 0.1 mm in diameter. The nucleus contains chromosomes carrying DNA. Ribosomes are molecular mechanisms making proteins. Lyosomes dispose of waste. Golgi bodies package proteins for transport about the cell and for secretion. Mitochondria provide a supply of energy by means of the Krebs Cycle shown in Figure 9.5 . (The shapes of components are not realistic.)

Let's imagine a trip into a living organism in which we use more and more powerful microscopes to examine its structure at ever finer and finer details. On the way down we will see the organism as a whole, then its organs, such as the liver, the skin, or the brain, and how they behave. Then we will see the structures of which those organs are composed – and at this level we are down near the level pictured in films such as *Fantastic Voyage* in which a submarine was shrunk to a size at which it could be inserted into a blood vessel and with its equally reduced human crew could voyage through the vessels and ducts of the body.[96] We are talking of things that are about 0.1 to 0.001 of a millimeter in size. We see individual cells, which, for all their variety, look not unlike the cells of independent organisms such as an amoeba or the plant cells of algae in a swimming pool (although of course animal cells do not have green chlorophyll in them).

We next look at the structure of the individual cells. Here we see great differences, because the cells of the different organs are specialized for different functions. Human red blood cells are relatively simple. They contain molecules of haemoglobin that gives blood its red colour; and haemoglobin is able to bind atoms of oxygen into cells when they pass through a region of high oxygen concentration in the lungs, and give it up to other tissues later. Many cells of the pancreas are specialized for producing the hormone insulin, and releasing it into the bloodstream to control the level of sugar being mobilized by the body to burn as fuel. Both red blood cells and the hormone-producing cells of the pancreas have rather short lifetimes, of the order of days, so that there is a constant death and renewal of these cells, with the processes of death and creation being nicely balanced in a fit organism. Cells are replaced by the division of existing cells, the materials of which are ultimately derived from our food, so that in a sense we are daily making living from non-living material. From before birth throughout the first decade of life new nerve cells develop and are added to the nervous system, and the number of mature and functioning nerve cells is thought to peak when a person is around age 20. From the twenties onwards they

---

[96] Currently *nano-engineering* is making mechanical devices at this scale, and these can already be inserted into the body.

die progressively, and are not replaced[97]. Nerve cells are specialized for transmitting electrical messages, and the long processes that run from one nerve cell to the next, the *axons*, are coated with fat which acts as an insulator and allows a very efficient messaging system in which pulses of electricity of several tens of millivolts[98] jump along the axon, beginning when the nerve cell is stimulated, and ending by stimulating the release of chemicals into the tiny gap between the end of the axon and the dendrites or body of the next nerve cell in the chain.

As we go down a further order of magnitude we find only chemical and mechanical characteristics of living cells. We are now too deep to see living things as such. At this level we see biochemistry, not biology. In all eukaryotic cells the mitochondria perform the chemical activities needed to make energy available to the cell; and the ribosomes, under the control of DNA programming, manage the synthesis of proteins. These activities are carried out by strange and very dynamic molecules, many of which move around carrying other chemicals to different places in the cell or through the cell wall. They are not themselves alive but are biochemical machines[99].

Chemical properties arise from the way in which atoms interact by exchanging electrons, a process that is well understood in physics, although the nature of chemicals can be extremely complex. On the one hand we find individual atoms and ions, the latter being atoms that have gained or lost electrons so that they carry an electric charge. For example, the electrical impulse in the nerve cell is caused by the difference in the relative concentrations inside and outside the nerve cell of sodium and potassium ions, Na^+ and K^+. Single atoms and ions such as Na^+ and K^+ are relatively simple structures, but organic molecules can be of immense complexity, now familiar

---

[97] Since, however, there are about 10^{10} nerve cells, 10 000 000 000 nerve cells, in the brain, you are not likely to end up with a brain rattling around in the skull like the pea in a referee's whistle, even though many thousands die every day. Recent research suggests that new nerve cells may form more frequently late in life than was thought in the past.

[98] 1 millivolt = 1/1000 volts.

[99] P. Hoffmann, 2012. *Life's Ratchet*

from accounts of DNA, the material which carries the hereditary programming of cells from one generation to the next, and which is a gigantic nucleic acid molecule, made up, in some species, of millions of atoms. Although the width of the DNA helix is only about 0.35 nanometres, an unfolded DNA molecule from a single human cell would have a length of over 20 metres, and can be packed into the cell only by very complex folding.

Large molecules, particularly the protein molecules, tend to be twisted and folded into elaborate three-dimensional structures. The DNA molecule is famously a double helix, coiled and recoiled on itself. Since like electrical charges repel one another and unlike charges attract, large molecules will tend to re-form themselves into a geometry dependent on the distribution of charges over the molecule[100]. So even in the absence of any imposed structure chemical components will organize themselves passively into characteristic three-dimensional shapes. Long molecules will tend to form three-dimensional folded ribbons. That will bring distant portions of themselves closer to one another, allowing in turn for even more complex foldings influenced by the electrical fields generated by the electrons and atomic nuclei. What is more, if we start with a particular molecule the folding will tend to be the same on different occasions, assisted by catalytic properties of enzymes. Fats and carbohydrates are simpler; sucrose, for example, being made of a rather simple set of carbon oxygen and hydrogen atoms bound together[101]. On the other hand, as the mechanico-chemical properties of molecules change, and some attract or repel others, completely new properties may emerge since many biochemical systems are self-organizing, catalysed by enzymes.

---

[100] There is a toy that consists of a collection of little magnets, each about a centimetre long and half a centimetre wide. When they are mixed into a disorganised heap, they spontaneously form into strings and lumps, because north poles attract south poles and repel north poles in magnets, just as do charges in molecules. This gives a feel for how structure can emerge from random collections of particles and forces.

[101] See Figure 9.2 below.

We know that the forces that bind atoms together into molecules are due to exchanges of electrons between atoms. For example two hydrogen atoms can use one electron each to bind to one atom of oxygen which can accept two electrons. The result is to make water, H_2O. The long chains of carbon atoms that give rise to complex carbohydrates and proteins can exist because carbon atoms can exchange up to 4 electrons. But these forces are not just randomly distributed. To take a very simple example, the way in which the electronic charges link the hydrogen and oxygen atoms in a water molecule means that there is always an angle of 104° between the H atoms in the molecule, and this in turn gives water certain physical properties, including the fact that ice floats rather than sinks, which probably have an important role in the origin of life[102]. The shapes of the big protein molecules mean that they fit together only in certain ways. There are often certain points on the molecules that make up the outer membrane of a cell onto which a virus particle, for example, can fit particularly effectively. This mechanical match allows the virus to connect itself to the cell and hence to pass its nucleic acids into the cell, infect it, and turn it into a factory for making more virus particles. In a similar way antibiotic drugs often attack infectious agents by mechanically locking onto the molecules of their cell membranes, or by occupying the sites on a cell onto which the infectious agent might fit. Enzymes often have geometric properties that favour the occurrence of certain chemical interactions between molecules.

A particularly delicate and subtle example of mechanics at the molecular level occurs in the nerve cell. Each nerve cell expends energy to pump Na^+ ions out of the cell, while K^+ ions diffuse into the cell, and it is the imbalance that causes there to be a voltage difference across the cell membrane of about 100 millivolts. When the nerve cell is stimulated by a chemical or electrical stimulus, the cell wall structure actually changes mechanically, and tiny channels open in

---

[102] For simplicity most of the descriptions of molecular properties will assume that atoms are fixed, deterministic structures, although as we have already seen, the more correct descriptions are in terms of probabilities of structural features in quantum theory.

the cell wall that allow the ions to pass through in a surge, so that the voltage difference is reversed for about a millisecond[103]. Perhaps the most amazing example of a mechanical system at the molecular level is the way in which flagella operate. These are the tails on motile cells such as sperm cells, or other flagellates in the world of bacteria and microorganisms found in ponds, soil, etc. When one watches such cells propelling themselves through a fluid by lashing their tails they appear to be waving them from side to side (of course, on a scale of less than 0.001 mm). But it turns out that in at least some species the tail, the flagellum, is actually turned by a rotary device like a motor: nature has evolved the original nanoengineering mechanical drive system, which at present is the only example of a "wheel" known to have evolved in biological systems[104,105]. In mitochondria parts of some molecules spin in response to electrical and magnetic fields at atomic dimensions to make available and transfer energy within cells.

The chemical and mechanical properties of living systems at this level are an endless source of fascination and wonder. The reason that living cells work is because of the chemical and mechanical functioning of the molecular, ionic, and atomic mechanisms interacting with one another, to provide energy, to transport chemicals, to cause motion (as in the sliding of molecules of actin and myosin when Saltarella jumps) and to give structural strength through chemical properties of bone. The important point is that as we examine a living organism at this level, as we examine the structure and function of living cells, we find only an enormous range of chemical activity, electrical activity, and mechanical activity. Many of the chemicals at this level of analysis, almost the last level above atomic physics, are unusual and normally found only in living organisms. But they are still chemicals. Whether we examine a single-cell organism or a human cell, all we find is an enormous range of chemical, electrical and mechanical activity. There seems to be nothing else to be found in living cells.

---

[103] en.wikipedia.org/wiki/Action_potential
[104] http://en.wikipedia.org/wiki/Flagellum
[105] P. Hoffmann. 2012. *Life's Ratchet.*

Of course all the chemicals the atoms, ions and molecules, do not float around in a disorganized soup. Within each cell there are smaller structures, organelles, of various kinds. We have already mentioned the nucleus, and there are in addition structures such as ribosomes and mitochondria. The mitochondria are highly organized organelles that contain a different kind of DNA from that which occurs in chromosomes, and mitochondria seem to be descended from bacteria that entered animal cells millions of years ago. Their metabolism benefits the life of the cell as a whole, and their own existence has become dependent on that of the cell they inhabit, a tightly symbiotic relationship. But the structure of these organelles, as indeed the structure of the cells as a whole, is entirely dependent on the electrical and mechanical properties of the atomic and molecular components. When the organelles such as ribosomes and mitochondria are needed to allow a cell to replicate, the way these organelles work depends on the shapes of their molecules, on the electro-mechanical fitting together of one molecule with another. At this scale it is the mechanical properties of molecules that play a central role in making a cell alive, although the molecules themselves are not alive but are merely chemicals. That is, if we look at each of them, we do not find that it is showing properties that we would describe as alive, although the cell as a whole does.

It is generally thought that living systems appeared on earth in a kind of primaeval soup, millions of years ago. Indeed bacteria and archaea forms of life may have appeared as long as 3 billion years ago, quite soon, in geological terms, after the earth formed. Some people find it hard to imagine how this could happen by chance, but we know that there are structures that will form mechanically in inert "soups" that would favour the development of cells. If we take a bottle with olive oil floating on water and shake it, the oil will tend to break up into droplets that float in the water, each taking a spherical form. These may merge to make bigger droplets, or can, in some liquids, spontaneously break up into smaller droplets, much as water drops falling on a windowpane may sometimes merge to make larger drops and sometimes break up into smaller drops[106]. Thus even before we have a well-organized cell with a nucleus we have the makings of a spontaneously self-organising system. Moreover, the oiliness of the drops will tend to attract and dissolve some kinds of chemicals but not others as we can see if we add a dye such as cochineal to our olive oil and water suspension. The oil droplets might even incorporate *within* them tiny water droplets if the latter had certain chemicals bound to them. Both water droplets and oil droplets often carry electrical charges, and so there will be an electrical field associated with them that will repel certain kinds of ions and attract others. The biologist Steve Rose[107] has speculated that such a process may be the way in which cells appeared in the first place, so that relatively simple self-replicating systems could have been present on the earth even before any living systems appeared. We know today that the cell membrane that bounds an individual cell in the watery environment of the body is made from lipid (fatty) molecules whose electromechanical properties provide a stable structure to contain the organelles and allow them to operate. We know that all the chemical elements that are required could have been present on the early earth, because we can model the origins of the solar system and its planets in the billions of years after the Big Bang. Some light elements[108] were formed as the initial radiation fireball cooled. The remaining heavier elements are formed in stars as they convert atoms into radiation, and are distributed through the universe by the explosions of supernovae. Amino acids, the compounds from which proteins are made, have been found in comets, and may have formed in deep space early in the history of the universe and so reached Earth.

Physical chemistry at the level of molecular interaction is effectively deterministic. If two entities carrying opposite electrical charges are close to one another they will be attracted and stick to one another. If they are close enough to exchange electrons new molecules will be formed according to their valencies, the number of electrons in the

---

[106] Another way to see these kinds of processes is to watch the changes in the coloured fluid in a "lava lamp".
[107] S. Rose. 2008. *Lifelines*. Lane Science
[108] That is, those with small atoms, mainly hydrogen, helium and lithium.

outermost shells of the atoms. If there is a membrane formed of lipid molecules with a high concentration of ions on one side and a low concentration on the other water will pass through the membrane to dilute the more concentrated solution. This determinism is modified by the effects of thermal energy. Molecules will be pushed randomly into contact with one another by the motion imparted by thermal energy. In short, if you put chemical species together in an appropriate environment they will behave in accordance with the laws of physical chemistry. If the components are of appropriate kinds they will form cell-like entities which will divide, and self-replicating entities may develop. In such self-replicating systems other electro-mechanical properties needed for the system to behave like a living system can emerge, generating new properties in accordance with the principles of emergence described by physics, complexity theory and biochemistry.

Today we already know how to analyse and synthesise DNA molecules. Indeed DNA has been transferred from one cell to another after the nucleus of the second had been removed, and with the insertion of new DNA replication occurred successfully. The species of a bacterium has been changed in this way. And it has been even possible to create a new form of DNA that has not existed before, insert it into a cell from which the nucleus had been removed, and it too has replicated.[109] So we are well on the way to synthesizing self-replicating systems in the laboratory. What has so far not been done is to make the cell itself and the organelles in it, such as mitochondria, ribosomes, and associated fluids. But we understand a huge amount about the molecular structure of the cell wall, in which lipid (fatty) molecules with a hydrophilic structure at one end and a hydrophobic structure at the other can make a hollow "waterproof" container. Furthermore we are rapidly learning about the molecular "robots", little autonomous protein machines that provide energy for the operations of the cell, store and transfer chemicals within the cell,

---

[109] C. Venter. 2011. *A life decoded*. London: Penguin Books. C. Venter 2013. *Life at the speed of light*. London: Little, Brown.

and and underlie cellular metabolism. A superb and very readable account of this research is provided by Hoffmann[110].

There is no doubt that soon artificial cells will be made. Whether or not one calls such entities "living" is a matter of choice. They will be composed of cell-like substances, take energy from their environment, fight against entropic degradation[111] and replicate. Note that to synthesise DNA is not to create life. DNA can exist as an inert powder, with no replicating properties: to replicate it must be in a cell – which will of course not replicate without the DNA.

In all our explorations of living tissue we find nothing but physics, chemistry and thermodynamics. Nowhere in the research has their been any kind of mysterious gap, nowhere any process that is at all unlike chemistry of carbon compounds as we know them from the study of organic chemistry. Nowhere is there a magical process that seems other than natural. The nature of living systems is mysterious in the sense that it is extremely complicated: it is not mysterious in the sense that it is supernatural. It seems reasonable to believe that it arose, in the five billion years of the existence of the Earth, "by chance", by random processes that brought together the elements, molecules and compounds that were sufficient to make self-replicating organisms which then evolved into the creatures we know today.

It is sometimes suggested that the events that could bring about living cells are so improbable that there cannot have been time for them to occur since the Earth cooled sufficiently for water to form. That is incorrect. We really have no way of estimating what the probabilities are for such a process, such a sequence of events. Furthermore you will remember that when we were discussing probability we saw that a rare event does not mean one that can only occur after a long time has passed, but one that does not happen often. In the case of the emergence of living systems once could be enough, and there may even have been local areas where the probability at times was very high, perhaps in association with hydrothermal vents, the "black

---

[110] P. Hoffmann. 2013. *Life's Ratchet*. New York: Basic Books.
[111] We will discuss entropy briefly below.

smokers" on the deep ocean floor,[112] or following a collision with a comet containing amino-acids. We don't know whether there have been many kinds of living systems of which those we see are the only kinds that survived. We do know that there seems to be almost no environment on the Earth as it is at present where life is not found. Bacteria are found in chemically hostile environments of concentrated alkalis and acids; in anaerobic environments at extreme temperature and pressure in the "black smoker" volcanic vents on the floors of the oceans; in Antarctic ice, and even within rocks deep inside the earth. Crude but remarkable laboratory experiments have shown that amino-acids and even simple proteins can form if chemicals thought to have been present in the atmosphere of the primitive earth, such as ammonia, acids, oxygen, etc. are mixed and subjected to the passage of electric sparks to provide energy. If we ask what the probability is that a collection of lipid molecules could have, by chance, formed a semi-permeable membrane, what kind of probability estimate are we making? We are talking of periods of time of the order of many millions of years, even billions of years. We are talking of molecules which are of the order of size of millionths of a millionth of a cubic metre randomly circulating in oceans that contain something of the order of millions of millions of cubic metres of water. There are more than 30 000 000 000 000 000 seconds in a billion years, so that encounters between molecules occur millions of millions of millions of millions of times in millions of millions of millions of millions of local environments. There is a constant bombardment by thermal energy stirring the molecules. It is meaningless to estimate the probability that a particular chain of physical chemical events will or will not occur at appropriate temperatures and pressures; but given the meaning of probability for rare events there is certainly no reason to think that it could not happen at least once!

It is interesting to see that the way in which living cells operate today depends on the interaction of chance and necessity. On the one hand are the deterministic electric forces that bind atoms together and underlie atomic and molecular interactions through the exchange of electrons and provide stability. On the other hand there is the

---

[112] See D. Attenborough, 2013. *First Life*. BBC DVD.

constant random motion of atoms and molecules, which we can see as Brownian Motion[113]. Without the determinism we would not have structure. Without the random thermal agitation the chances of molecules getting close enough to one another to interact would be greatly reduced: we would not have the dynamics. Chance actually supports the deterministic activities of life[114].

Faced with these kinds of numbers, and the kinds of events needed for living systems to emerge, the quantitative aspects of probability fade away, and one is left with an intuitive decision, almost an aesthetic feeling, about whether or not it could happen. Given that we know there are living systems around us, and given that when we analyse them we find only chemistry and physics in the processes that keep them in being, it seems not unreasonable that living systems could arise by chance. Similarly, it seems to me that calculations about the probability of life on other planets around other stars, the so-called *Drake Equation*, are simply silly as a quantitative exercise. There are unimaginable numbers of stars with enormous numbers of planets in our immense universe. If living systems arise by chance when appropriate molecules come into contact, it clearly could perhaps happen somewhere[115]. Trying to put numbers on the probability of the emergence of living systems is simply futile; but the Scientific Story about the origins and analysis of life on earth are I think convincing. No transcendental magic is needed. It may even be that by the time this book is in print scientists will claim that a complete living system has been synthesized in the laboratory.

A Philosophy of Life

If science is straightforward in what it has to say, what about philosophy? What about the traditions of belief that there is a special change in matter than makes a living system from a non-living one,

---

[113] If pollen grains are suspended in water and viewed under a microscope they constantly jiggle randomly because they are bumped by the random movements of water molecules. This is called Brownian Motion.

[114] P.Hoffmann. 2013. *Life's Ratchet*. New York: Basic Books

[115] In addition to here on Earth, that is.

and that at death the process is reversed? What about the *souls* of living things? The discussion of life that we have just followed is very reductionist. It tends towards a "nothing-buttery" account of what it is to be alive, namely, that living systems are made up of "nothing but" certain kinds of chemicals, atoms and molecules in certain combinations; and that if a thing is made up in that way then it will be alive. That is no doubt true, but the story of the "advertisement" warned us to beware of "nothing-buttery". Nothing-buttery may be a prescription for a healthy diet, and in discussing explanations may make a story simple, but it may also be misleading. Think again of Saltarella. She was "nothing but" the molecules of which her body was made, but an account of why she jumped is unable to capture the reality of the jump with a "nothing but physics" account. So what should we say as we come up the ladder of description about living systems? Will we find that we have to talk about *souls*? So far the answer seems to be, "no".

We will look at this topic in detail in the final chapter of the book, but a few words are in order here. To start with, let's consider how the idea of a soul might enrich a reductionist Story, even if we do not want to talk about GIMs. Even for a reductionist account we need a richer language. At least we need to talk, as we did at the start of this chapter, about different kinds of cells, which form different kinds of organs, which have different roles in a complex living organism such as a plant or a human. It is indeed not possible even to give a reductionist account of the nature of a living system in terms of the DNA which causes the species characteristics of a living thing, because between the formation of the zygote, the first cell in the history of a new individual formed by the fusion of the male and female gametes, and the fully grown organism, there are events that are not only due to DNA programming. DNA itself is not alive, but just an inert nucleic acid molecule. The fact that we can talk about pluripotent *stem cells*, which although having identical DNA will in different chemical environments develop in different ways, shows that we need a richer description. Indeed we may well find that it is again useful to go back to the different kinds of causes that we saw were introduced by Aristotle. Let's see how such a higher level description of a living system might emerge.

Kidney cells are programmed in the DNA of a mammal, and develop from their biochemical precursors under the programming of DNA and the local environment in which stem cells find themselves. Why do kidney cells develop, and indeed why do kidneys develop? Well, in a very straightforward sense, in order to provide a way of filtering out certain waste products from the body and to maintain its chemical state. The *final cause* of a kidney is to provide filtration of the blood, to filter out undesirable and toxic chemicals, and to preserve the fluid balance of the body. The *efficient cause* of a kidney is the interaction of the DNA programming with the environmental pressures of the body chemistry on the stem cells. The *material cause* is the biochemical composition of the kidney cells. Because all three kinds of causality are active in the living body the result is a functioning kidney playing its role in the urinary system of the body. An Aristotelian analysis of cause applied to the biology of the kidney lets us translate the lower level reductionist story into a higher level purposeful developmental story about how the mature body functions. This story will only be true as long as the organism is alive *as a whole*. Under laboratory conditions we can keep an organ, such as a kidney, alive for a considerable length of time, as is done during kidney transplant operations. But this of course does not mean we are keeping the original owner of the kidney alive. Ultimately a kidney is only a kidney when it is in a body and performing its function. Indeed what happens when an organism dies is that the body loses its integrity, and the organs begin to act as individual living components rather than as a single animal. After a while this cannot continue, because the overall integrated functioning of the animal ceases, including the intake of food and oxygen, the manufacture and circulation of energy-producing chemicals, and the excretion of waste gases and fluids. The organism dies, and then its parts, its organs, die. By "dies" here we mean, "stops performing the full range of activities it performs when part of a whole organism". In the end, it reverts to its chemical components which as chemicals return to the environment, available for further chemical interactions as chemicals, not immediately as living cells.[116] It is this fact of the

---

[116] "Dust to dust, ashes to ashes: Into her tomb the great Queen dashes." Indian poet on the death of Queen Victoria.

organism existing and acting as an integrated whole that Aristotle defined as the meaning of *soul,* a way of speaking that is clearly useful and meaningful, even though it makes no reference to spirituality or immortality.

The struggle to maintain the integrity of a living system at every level, cell by cell, organ by organ, and as a single whole organism is perhaps *the* defining characteristic of living systems. Although it is not directly perceptible, this spontaneous and natural resistance to increasing entropy may be the best definition of what it is to be alive. *Entropy* is a measure of how well organized a system, any system, is. The more order that is present, the less the entropy. The natural thermodynamic processes of the universe tend always to increase entropy. As time passes, systems become degraded and disorganized, and energy tends to become uniformly spread out over larger areas of the universe rather than being contained in a small pocket such as an organism. While an animal is alive it is highly organized. It is recognizable as a particular individual. Its body is composed of identifiable organs with identifiable functions; and each organ is composed of identifiable kinds of cells performing their functions to maintain the function of the organ. The biochemical functions of a cell maintain its integrity even in the face of hostile changes in its immediate environment, changes in acidity, in oxygen concentration, in energy supplies. Only at death do all these levels of organisation disappear, and the components and energy of the organism become increasingly randomly spread about the environment. The body rots and its chemicals diffuse into the environment: "ashes to ashes, dust to dust". But during life, the organism retains its integrity and functions as an individual[117]. It is, as Aristotle said, ensouled.

Now the reductionist description that we gave of living systems does not capture this notion of the integrity of the individual that is composed of many different chemical processes and organs. Indeed in none of our descriptions so far does this notion of the identifiable

---

[117] There are some interesting organisms where this is hard to see, such as certain kinds of slime-moulds; and it is not clear whether a virus is best described like this. But then of course there is doubt as to whether viruses are living.

individual as an integrated dynamic organism resisting entropy appear. If at all, it is in this aspect of living system that the notion, the Philosophical Story, of a *soul* may be useful. But we have to be careful, because very often in Everyday Stories people take it for granted that *souls* are parts of a person that are not material, and we have already seen, in an earlier chapter, that there is really no reason for such an idea. If however a *soul* is not an immaterial mechanism but rather a principle of organisation, then indeed it is not a *thing*, neither physical nor non-physical: principles are not things, even when they are causes. Even when we discussed who the *real person* is, we found that we should not invoke the notion of an immaterial part of a human, a GIM; how much less in the physical account of being alive?

So to summarise, we can give an account of what it is to be alive without any appeal to a non-physical agent that somehow adds magical properties to chemistry. But at the same time, we need an account also of the fact that the myriad chemical activities produce identifiable individuals that resist entropy. That is what the Story of the *soul* seems to have been originally about, at least for Aristotle. It looks as though we can expect to synthesise life, in the sense of making a living cell, before too long. So for now, let's turn our attention to how from the original simple cells, in the course of millions of years, new and more complex kinds of organisms developed, and eventually produced humans.

Chapter 8

Evolution

> If he should dress respectfully, and let his whiskers grow,
> How like this Big Baboon would be to Mister So-and-so.
>
> Hilaire Belloc. *A Bad Child's Book of Beasts.*

Why Things Evolve

Scientists may soon synthesise living from non-living materials. Of course they would first make a "simple" cell, not a complex adult animal, much less an adult human. And if indeed living systems appeared spontaneously we can assume that at some time in the past the world had only such simple cells, perhaps something like bacteria, algae, or amoebae. How did we get from that state to today's richness of plant and animal life? How did we get from single cells to humans? The Scientific Story to account for this is, of course, *evolution*.

Many people are upset by the idea of evolution because they assume that that it rules out Religious Stories about human life. That is true if one's Religious Story says that the world is only a few thousand years old rather than, as science suggests, many millions of years old and located near a star that is about half way through the 10 billion years or so for which it is expected to exist. If on the other hand one is open minded about how old the earth is, evolution could perfectly well be the way in which a god (should one exist) brings humans into existence. Our belief that the earth is some five billion years old is not arbitrary. It is based on an analysis of the physical and chemical properties of the rocks that make up the earth, the rate of radioactive decay, and other evidence. And remember that we saw in Chapter 4 that these are not just isolated facts, but that the physics that supports this analysis is the same physics that allows us to make everyday things such as mobile telephones, computers and aircraft. If the science that gives us an estimate of the age of the earth is wrong, then these everyday devices should not work – but they do.

Evolution is usually thought of as a theory about how biological species change, but it has a more general meaning. It is a theory about how any complex self-reproducing system changes its properties autonomously. There are several theories of biological evolution, Darwinism, neo-Darwinism, Lamarkism etc., and variants of each of these. To understand classical Darwinism the best and most readable book without doubt is Darwin's own *"On the Origin of Species"*, although Dawkins's earlier books *"Climbing Mount Improbable"* and *"The Blind Watchmaker"* are also highly readable[118]. We won't try to decide here among the various versions of evolutionary theory. But we will try to understand the basic ideas, and to see some evidence for the existence of evolution and the nature of evolution *as such* (rather than a particular version).

What is "Evolution"?

Not only biological systems evolve. Evolution is a more general concept. We say that a system evolves if it is composed of entities that reproduce themselves (that is, replicate) and whose properties change so as to increase the probability that the system will continue to exist in the environment it inhabits, without any intervention by an outside purposive agency. Repeated replication means that desirable characteristics become more strongly displayed, and undesirable characteristics disappear. It is common (but not necessary) to call such a system a population, and in biology the "desirable characteristics" are those that increase the probability of reproduction.

---

[118] C.Darwin. 2008. *On the Origin of Species*. Oxford. Oxford World Classics.

R.Dawkins. 1996. *Climbing Mount Improbable*. London: Penguin Books.

R.Dawkins. 1988. *The Blind Watchmaker*. London: Penguin Books.

It should not be a surprise that some systems evolve. Rather, it would be very surprising if any replicating system did not evolve. Evolution is the usual way for replicating systems to behave. There is nothing special or particularly startling about evolution. Let's see why.

Here are the minimal properties needed for a system made up of many individuals to evolve:

1. Individuals replicate if they survive long enough.
2. Replication is not exact: offspring are at least slightly different from the parents.
3. There are properties that the individuals possess that are needed for survival. We call these desirable characteristics.
4. Individuals are sufficiently similar that they compete for resources in the environment.
5. The environment selects characteristics in each individual in the sense that some individuals are fitter than others to use the resources of the environment and hence reproduce.
6. The environment favours individuals that have more of the desirable characteristics, in the sense that the latter make them more likely to replicate.
7. To make the description very general, let this variation from one generation to another be random, (it happens by chance,) both as regards which offspring are affected, how and to what extent.

Given these properties a system *must* evolve, because those individuals that have more of the desirable characteristics will be more likely to reproduce, passing on also any other characteristics they possess. It is very important to remember that when we use words like "favours", "needs", "selects", "replicate themselves" and so on, we do not mean that any part of the process is conscious, purposive, or has deliberate goals. When we say, for example, that an entity "needs" to survive to an age where it can reproduce, we simply mean that if it does not reach the age of replication no offspring will appear. When we say that the environment "favours" or "selects" a characteristic all we mean is that if an entity possesses that characteristic the probability of replicating and passing on its characteristics to its progeny will increase.

Here is an example. A population of Peppered Moths lived in the North of England on the trunks of trees in the early 19th century. The moths were a sort of "pepper and salt" colour, predominately white with small black spots, and birds preyed on them. The moths reproduced, but not all members of the population were exactly the same. Some were more "peppery" (darker in colour), and some were relatively "salty" (lighter in colour). The trunks of the trees were also a sort of greyish "pepper and salt" colour, so the moths were, on the whole, pretty well camouflaged, particularly the lighter ones. When the Industrial Revolution occurred, the trunks of the trees became blackened by soot. Now the property of "matching the colour of the tree trunk" became important, and the strength of this property was "favoured by the environment", the measure being the ease with which the predators (birds) could find the moths and eat them. Clearly those moths that were more "peppery" had a better chance of surviving until they could breed. By the mid-20th century the population of moths was breeding a majority of very dark moths. When the Clean Air Act came in the mid-20th century, the trunks of the trees became lighter and lighter over a period of years – and so did the moths.

Here is an account of the changes in the moths taken from a recent newspaper report[119].

> Peppered Moths: Moth turns from black to white as Britain's polluted skies change colour.
>
> The moth was white with small black speckles but over time it evolved to being almost black in parts of the UK because of heavy industrial pollution. The change made it less obvious to predators against backgrounds of grime and soot. Having declined by more than two thirds compared to 40 years ago, it is regarded as a classic example of natural selection and has consequently become known as "Darwin's moth." Now in post-industrial Britain, 200 years after Darwin's birth, the moth is changing back to its

---

[119] Published: 12:57PM BST 19 Jun 2009 DAILY TELEGRAPH

original white colour. Scientists at Butterfly Conservation, based in Dorset, are now appealing to the public for help in finding out how widespread this change has become. As part of Garden Moths Count 2009 they want people to search their gardens for the moth and log their sightings.

"We have seen these moths making a big swing back to their original colour," said Richard Fox, project manager of Moths Count. "It has been happening for decades as air pollution is cleaned up and with the demise of heavy industry in the big cities. The moths have been responding to this and the numbers of black and white moths will vary across the county. In Dorset it is very rare to see the moth in its dark form, but in industrial cities 150 years ago they were almost all black and that's where we will notice the greatest changes now."

That is what we mean by evolution as a result of natural selection. It certainly happens.

Evolution can occur even in non-living systems. For example computer scientists took forty years to develop reasonably effective programs for speech recognition and optical character recognition to scan written text into computers. One approach to this problem was to use what is called a *"genetic algorithm"*. The material to be recognised is processed so that it is suitable as input to a computer. The program is given a criterion for what counts as a successful outcome. Someone writes a program that analyses the input into several components giving each a numerical weighting, (a numerical index of its importance) and tries to use them to identify the pattern. The programmer tells the program which of its attempts are the more successful, and the program stores those weights, and then continues randomly varying others as it is given new inputs. These variations are equivalent to mutations in biology. Here replication consists of the program being rewritten with its changed weights, and randomly varying some weights in each generation. The property to be optimised, the "desirable characteristic", is the score on recognition as measured by the human programmer. The program thus evolves new ways of analysing inputs. Note that here the only population is a collection of programs; the variation during replication is random; and the human is the cause of the evolutionary selection, "preying on", "killing off", programs that do not do well. But in these systems evolution of pattern recognition certainly occurs. The human programmer in the end may not even understand how the program is doing the task: the algorithm has evolved by itself.

So there is nothing particularly exciting about evolution as such. Any system that has replication with variation and that is subject to selection *must* evolve: it can't do anything else! That is true for biological systems, computer systems, and would even apply, (if they reproduce with variation and have environmentally selectable characteristics) to ghosts or angels. In no case is intelligent intervention by goal directed intention needed. Random changes and environmental pressure are sufficient.

"On the Origin of Species"

So why do people get so upset about the idea of evolution? Darwin suggested that evolution by means of natural selection is how animal and plant species arise and change, and how eventually humans appear. What I have called the "important property" that is favoured by the environment is "staying alive long enough to breed the next generation". The process is completely random and purposeless, and what looks like goal directed development is actually due to the chance relation between random changes in the organism and the physical characteristics of the ecosystem. The fact that the members of a population become more likely to survive and breed is not "for the good of the species" but merely a chance accident. It is we, looking at the events, who choose to use words such as "for the good of the species": nature is blind and indifferent.

At the time Darwin was writing many people believed that all species had been made in their present form by God as described in the book of Genesis, and in particular that Man was made in his present form by God. Hence the idea that one species could change into another, and in particular the idea that Man had evolved from other species, was unacceptable to many. The crucial claim in this chapter, by contrast, is that *usually* one species changes into another. Indeed such changes are constantly present in the ecosystem of the world (and even, as we have seen, in some non-living systems). The example of the moths shows that there can be great changes in the physical form of an animal. This will almost always mean that its behaviour will also change, since a change in physical form however slight offers new possibilities of how to behave, removes others, and opens the door to the impact of selection. To bring this home I will talk briefly about some of my own research.

An example of the evolution of behaviour.

To show that evolution can select for behaviour let's look at an experiment that was carried out in the 1960s in my laboratory at the University of Sheffield. We used the fruit-fly, *Drosophila melanogaster,* and set out to see whether we could produce a subspecies that had different behaviour from wild flies. The experiment was reported in *Nature* in the 1960s[120].

We started with a population of "wild-type" flies, that is a population with no special properties, such as you might find on a fruit stall in any street market, and divided it into two groups of several hundred flies. One group lived and reproduced in an environment with normal food, the other in an environment with food adulterated with peppermint oil, which flies usually avoid if possible. After several generations those forced to live in the adulterated environment changed and became a population that instead of avoiding it would seek out peppermint flavored food. Those that lived on normal food did not change. When after several generations we took the "peppermint" population and put it back into the normal environment, the preferences of the flies did not immediately change back. We had converted a normal wild-type population into a population that liked and chose to feed on peppermint-scented food. Behavioural evolution had occurred. The experiment took about a year, and some 12 generations. We repeated the experiment three times, starting from

---

[120] Moray, N., and K. J. Connolly, "A Possible Case of the Genetic Assimilation of Behaviour," *Nature*, 199, 353-359, 1963.

different wild populations. Our intervention, forcing the flies to live in an adulterated environment, is comparable to a wild population in an environment that has undergone some major ecological change. We were the equivalent of the industrial soot in the moth case.

Can a new species be created by evolution?

Were the black moths a new species that differed from the pepper-and-salt moths, and are the white moths yet another species? Were the pro-peppermint flies a different species from the wild-type flies?

The concept of "species" is not as simple as it seems. Certainly we would all agree that a mosquito is not the same species as a giraffe. But is, for example, a poodle a different species from a spaniel? A traditional definition of species used by biologists was that individuals of the same species can interbreed and produce fertile offspring. So poodles and spaniels are the same species, (my step-daughter has a "Spoodle" bitch which is fertile): but horses and donkeys are different species (because mules are infertile). Today there are known to be exceptions to this definition for both plants and animals, and a discussion of species is usually related to the properties of DNA in different animal, and to the way in which creatures breed true and alter in cross-breeding. The definition of *species* is not self-evident.

Given natural and laboratory experiments such as those just described, is there any reason to reject the idea that in principle one species can change into another? We certainly know that species can disappear. No dinosaurs now wander the Earth. Closer to our time are the disappearances of the Dodo in the 19th century, and of the Passenger Pigeon in the 20th century. Of course no one would suggest that one can change a mosquito into a giraffe. What we would expect to find, described very beautifully by Dawkins in *Climbing Mount Improbable*, is a series of very small changes in form, behaviour, and biochemistry, each almost trivial by comparison with the immediately preceding stage, but which at the end lead to animals at the extremes of a series that are very different, so different that while any two immediately adjacent members of the series would be inter-fertile, those widely separated would not be. Adjacent members of the series would not be different species. The end-of-series members would be different species.

In biological evolution it is changes in biochemistry that are crucial. They underlie changes in anatomy and function. It could be that the structure and overt behaviour of successive generations of animals are not very different, but some chemical change occurs that makes one animal's spermatozoon induce an anti-body reaction in the other, resulting in infertility. On the other hand, a mutation in a single animal can result in many animals being born with that difference in the next generation, so the new version can spread into the population. There seems no reason why such an evolutionary process could not produce new species. And if the evolution story is not correct we still have to account for the fact that at one time there were no mammals according to the fossil record: now there are mammals.

To accept that evolution produces new species says nothing about whether the entire system needs the addition of a Religious Story about "intelligent" cause and effect, merely that the accumulation over many generations of small changes each occurring by pure chance is sufficient cause. Nor does it say that "spiritual" or "transcendental" properties could not arise: but that of course would depend on our definition of such words. Animal behaviour other than that in the simple Protozoa requires a nervous system, and nervous systems are biological systems that can evolve. Remember again that the science that says that the world has been around for a very long time (not just 6000 years), and the science that proposes the theory of evolution, is the science that makes possible your everyday gadgets such as mobile phones. Furthermore we now know that evolutionary changes can be very rapid[121].

Where are the intermediate forms?

Some people ask why if there is a continuous change do we not find intermediate forms in the fossil record? Well, we *do* find many intermediate forms, but they are not close enough to those at either

---

[121] J. Weiner. 1995. *The Beak of the Finch.* New York. Vintage Books.

Science, Cells And Souls

end to satisfy some sceptics. Let's think about the nature of the scientific evidence, and attitudes to it.

Suppose we find two types of fossils, which are similar to each other but are very different sizes. I say I think that the big ones evolved from the small ones: after all, the big ones are in a different and much more recent geological stratum, and apart from their size closely resemble the earlier ones. You say that can't be true because there are no fossils of intermediate size in intermediate strata. I then find some middle-sized fossils in middle-aged strata. But you then say, "Ha! But you have not found any between the middle sized ones and the big ones!". And then I find.....but you say..... and so we go on. Clearly this is not the way to conduct the discussion. You can always say that the remaining gap is not bridged by any fossil found so far: I can always say that somewhere there may be another to fill that gap. As is often said, "Absence of evidence is not evidence of absence." The chance of any particular individual animal being fossilised is extremely small. There must be millions of species for which we have no fossil remains, and certainly there are billions or trillions or more individual animals that died and were not fossilised, and which may have included the particular form that you are demanding if you are to accept my story of evolution.

The Dusky Paradox: a fable.

Someone once said, "Night is so different from day that one cannot come from the other." Obviously, at dusk there is no moment when one can say, "A moment ago it was daylight; now it is night." But there is no doubt that night is different from day. The change from day to night is not like the throwing of a switch to darken a room, but by nuances of brightness during dusk, so that while adjacent moments differ imperceptibly, the end points are lit respectively by the sun on the one hand and the stars on the other.

Your objection is like the Dusky Paradox. You are saying, "I don't believe that night came from day. Night is too dark. How can it come from the bright day? Where is the intermediate stage?" And I say, "Well, at 6 o'clock it was bright, and at 10 o'clock it was dark. Look at this photo taken at 8 o'clock. It is half way between." And you reply, "Ah! But you have not shown me how the 8 o'clock brightness changed to the 10 o'clock darkness." And our argument goes on. To consider the Dusky Paradox shows how illogical is the objection from gaps in the fossil record.

One series for which we have a very good evolutionary series is the Horse, with a large number of intermediate fossil "horses" of different shapes and sizes forming (for me) an overwhelmingly convincing series indicating slow evolution over millions of years from small dog-sized pre-horses (Eohippus) to today's horses. Another is the evolution of Whales (Cetacea)[122]. As a very different example, let me again refer to some of my own research. This time it is work on marine copepods, very small animals rather like waterfleas. I worked with R. L.Gregory on the visual system of *Copilia* and related animals in the 1960s. We were interested in a species, *Copilia quadrata* that has a very striking eye, as shown in Figure 8.1. It resembles the compound eye of an insect, but it has only one photoreceptor on each side of the body. The front of the body has a large and very beautiful lens on each side of the body, and behind it, half way along the body, there is a single photoreceptor structure (a rhabdom) on each side. The remarkable thing is that the animal seems to use a muscle to sweep the receptor across the field of view in the focal plane of the front lens, as if scanning the image formed by the front lens. While there are problems with that interpretation, one can show that this scanning motion is a more effective eye for detecting nearby objects entering the visual field than a fixed eye even with several receptors[123]. We can therefore interpret it as the evolution of a mechanism to improve vision.

It was exciting to also find several species of *Copilia* and other more distantly related copepods with eyes that were different from each other, but which, taken together, look very much like a set of solutions

---

[122] http://www.darwiniana.org/landtosea.htm
[123] Gregory, R. L., N. Moray, and H. Ross, "The curious eye of *Copilia*," *Nature*, 201, 116-1168, 1965. Moray, N., "Visual mechanisms in the copepod *Copilia*," *Perception*, 1, 193-207, 1972.

due to natural selection for gradually improving vision. See Figure 8.2. In *Copilia* males and some other species there are just lenses on

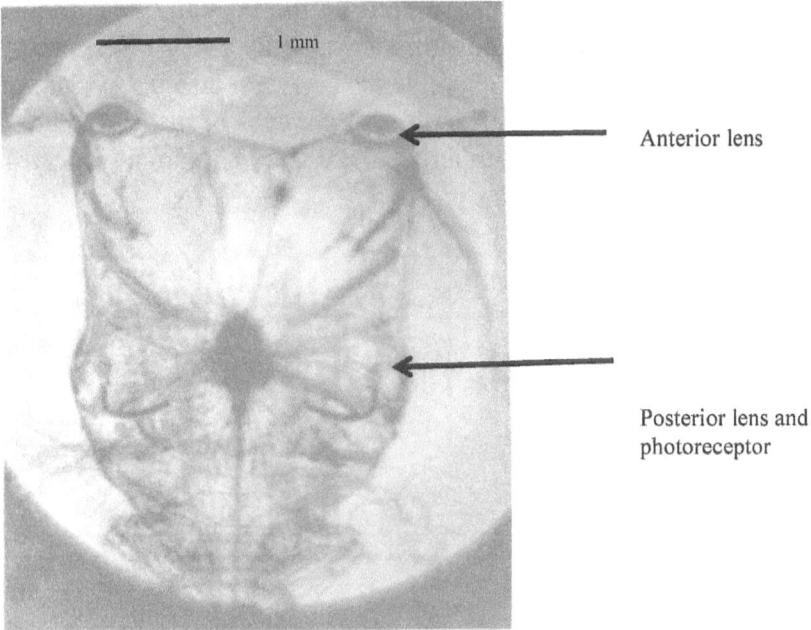

Figure 8.1 *Copilia quadrata*.

a central single yellow structure that contains the light-sensitive cells. In *Nesippus* the visual receptor structure is Y-shaped with a lens on the end of each arm of the Y. In *Copilia mediterranea* there are two straight receptor structures side by side with lenses on the front of the animal and crystalline cones on the receptor structures. In *C. denticulata* the two structures are more widely separated, and are partly bent. In these two species of *Copilia* the receptor structures are static. Finally in *C. quadrata* the receptors are widely separated, are bent into a right angle, and are scanned across the focal plane of the front lenses by a sheet of muscle. Overall they provide an example of different current intermediate forms, each representing a solution to the problem of vision, and together showing how slight, "partly developed" organs would give a considerable improvement in function even though a single end point on the path from a simple primitive eye has not been reached[124]. The argument that these different forms are all related to the evolution of improved visual acuity is supported by quantitative calculations that show that one would need several static receptors to be as good as one moving one – and there seem to be no animals with only a very few receptors. Water-fleas (*Daphnia*) for example have several dozen. So there *are* examples of series of small but useful changes of just the kind one would expect if evolution underlies the differences among species.

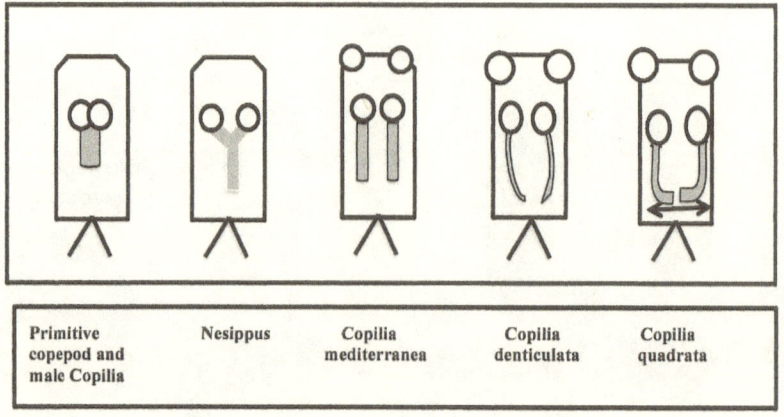

| Primitive copepod and male Copilia | Nesippus | Copilia mediterranea | Copilia denticulata | Copilia quadrata |

Figure 8.2 Species of copepods suggesting changes in the evolution of the visual system which provide ways of improving vision.

Let's go back to the problem of intermediary forms. If anyone wants to challenge the evolution story and to say there are gaps in the fossil record, then they should also say what, for them, would count as filling the gap. Suppose it is a question of height. If we have one fossil 3.0 metres tall, and another 0.5 metres tall, I would agree that intermediate forms seem to be missing. But how fine a series do you need to be convinced? Will 0.5, 2.0, 3.0 metres do? Or do you want a series 0.5, 0.51, 0.52, ... 2.8, 2.9, 3.0 metres ? Or even, 0.500, 0.501, 0.502, ...metres? If you choose the latter two series I can be sure that I will never find them, not because there may not have existed

---

[124] It is not the case, obviously, that evolution requires the new organ to appear fully developed. See R. Dawkins, *Climbing Mount Improbable*.

individual animals with those heights, but because the probability of finding (in the last case) 3000 different fossils of just those sizes in the series is near zero, even if the animals once existed. By choosing one of the last two series what you are really saying is, "Since I don't believe there is such a continuous series, I will set a criterion that I know you can never meet; so I cannot be proved wrong." (As we saw when we discussed why we believe in science, in order to have a discussion about the existence or non-existence of something, you have to say what your criteria are for accepting the case you do not believe in; otherwise there cannot be a discussion. And even then, if you want a genuine discussion, you must not set criteria that will always protect your belief.)

Why don't older species disappear?

Is it evidence against evolution if new animals appear but the old ones don't die out? In fact that is what one would expect. The evolutionary changes are made with respect to some "ecological niche" currently occupied by an animal or plant. Consider the case of goats living in a desert environment where there is little grass but many small trees and bushes. The hooves of normal goats have evolved for walking on more or less flat ground. But suppose that in some locations there are low trees or bushes with succulent leaves, but only grass in other locations. Any goats whose hooves were a slightly different shape that allowed them to climb well in the branches would be better nourished and more likely to breed successfully. The goats with modified hooves might well over many generations be favoured in the environment with trees, and might eventually end up with seriously modified hooves that were better at gripping branches. At the same time that they evolved to occupy the "low tree ecological niche" where they had a biological advantage over goats with unmodified hooves, the original goats and their descendants would still be living albeit with difficulty on the grasslands. In terms of survival to breeding age the "climbers" might not be able to run as fast on the flat land, while the "flatties" would be faster but not be able to climb to reach food. In fact there might be more food for those who remained on the ground because their competitors had taken to the trees. So evolution would predict that the old and new model goats could co-exist. It would need some really catastrophic change in the environment, one that abolished the original ecological niche, to make the original type of goat disappear. Dramatic examples of the interaction between animals and their ecological niches is shown in the beautiful work recently done on Darwin's Finches[125].

The relation of structure and function across species is also wonderfully illustrated by the biochemistry of genetics. There is a gene, called Pax-6, that is responsible for the development of eyes in insects. If you introduce an extra Pax-6 gene in an insect larva it develops a third eye. The Pax-6 gene also occurs in mammals, surprisingly, where it causes a *mammalian* eye to develop. Amazingly if you take a Pax-6 gene from a mammalian chromosome, and put it into an insect cell, the insect develops a third eye. But *although the gene was taken from a mammal where it causes a mammalian eye to develop, when put into the insect the same gene, Pax-6, causes an insect eye to develop.* Marvellous! Incidentally, this also shows that genetic determinism is false. Although eyes are caused to develop by the gene (a deterministic effect), the way in which the gene is expressed depends on the biochemical environment in which it finds itself.

Think what this implies for cloning. If the two members of a cloned pair of human cells were fed on different diets, that would cause significantly different biochemical environments during development, and hence different developmental paths. Such differences would be increased by the impact in later years of education, training, opportunities for different behaviour patterns, etc.. The notion that cloning produces deterministically indistinguishable identical copies is obviously false. If we cloned a human, by definition from a single cell, and the original person grew up in an English family while the clone grew up in a French family in Niger, would you really expect them to be indistinguishable? Clones are just identical twins born at different times. Since specific memories are not present in the zygote but are added to the nervous system by experience, and the nervous system of an individual is the result of interaction between

---

[125] J. Weiner. *The Beak of the Finch,* 1995. New York. Vintage Books.

the genome, biochemistry, and environmental factors including education and socialization, cloning from a single or a few cells would certainly not transmit all a person's memories, let alone the complete personality. A clone from a single cell cannot be, by the time it has become an adult, an exact replica of the individual adult from which the cloned cell was taken. Suppose I clone myself from one of my cells. Obviously the clone will not be identical to myself, since it will not have my memories and will grow up in a world with a completely different set of exeriences, different history, different diet, etc..

Figure 8.3 Goats climbing trees in Morocco. Are their hooves becoming more claw-like?

There is no doubt that natural selection operates. For a brilliant, exciting and indeed romantic account of the scientific study of evolution in our time, read the wonderful book by Weiner already cited. He describes the work by two scientists, Peter and Rosemary Grant, on the biology of Darwin's Finches on one of the Galapagos Islands. Tiny changes in the size and shape of the beaks of these birds, changes of less than a millimeter, interact with changes in the vegetation to produce evolutionary changes within just a year or two, changes in behaviour, in physical form, and in the way the different species respond to changing ecological niches. Weiner describes similar rapid evolution in fish, and the way in which antibiotics cause evolution in bacteria to produce resistance to treatment within a few hours of exposure to medication. And all these examples show changes caused by natural selection not over millions of years of so-called "geological time", but well within the lifetimes of you or me. Evolution, as I saw in my laboratory in the 1960s, can be very rapid and readily observed.

There is no doubt that evolution occurs. The logic of an evolving system makes it certain to occur when we have replication in a varying and demanding environment; and furthermore there are experimental examples where it happens. As to whether evolution is a complementary or an alternative account of the natural world to that given by religion, it depends on your religion. I see no reason why a religious person should not say that God works in evolution his wonders to perform. I just happen to think that such a description is unnecessary.

Chapter 9

Making a Person

What are little boys made of?

Slugs and snails and puppy-dogs' tails, that's what little boys are made of.

What are little girls made of?

Sugar and spice and all things nice, that's what little girls are made of.

(Traditional)

Suppose we didn't want to wait for millions of years for evolution to produce humans. Suppose that, like a medieval alchemist, we wanted to make a human. Could we do so? Could we make a person, not as a clone, but from scratch?

The physics and chemistry of life

We are live and active creatures. We take in information through our senses, interpret the state of the world, feel anger and love, make decisions, and interact with our environment (which includes other people). Do we therefore expect to find something unusual, even magical, in the composition of humans? Stones can't do these things, nor can rhubarb. Horses can't do many of them, and not even cetecea or primates such as chimpanzees have language like ours. It has been said jokingly that culture is all those things that we do that monkeys can't do. If we can do these things what are we made of? What if anything is special about a human? We take up the story we left at the end of the Chapter 7. Even if we could make a living cell could it be a human cell? Let's start by asking whether we could make a single cell, from which a person could develop. That after all, is what happens in everyday life.

Bottom-up biochemistry

If we examine a person "scientifically", what will we find? Let's again imagine an instrument like a microscope, and peer down into the structure and function of a human. Let's pretend that it is so powerful that with it we can even see individual atoms, and that we can do "nanochemistry"[126] at those levels. At the deepest level of analysis we will find electrons, protons, neutrons and their component parts (such as quarks). But these do not reveal anything about human nature, because all electrons are alike, as are all protons and all neutrons. When we examine a subatomic particle such as a proton we cannot tell what creature we are looking at, indeed not even what substance, what chemical element, we are examining. There is nothing special about the subatomic particles that make up people, because there is nothing special about any subatomic particle. Let's back out a little, and look at atoms, that is *collections* of subatomic particles. Now we can see what chemical element is present. If we find combinations of just one proton and one electron bound together, we know we are looking at a hydrogen atom. If we find two protons, two neutrons, and two electrons bound together, we know we are looking at helium. If we find six protons, six neutrons, and six electrons, we know we are looking at a carbon atom[127].

We will certainly find all of these in a human, but it still does not tell us very much, because all atoms of hydrogen, (denoted as **H** by

---

[126] "nano" means 1/1000 000 000 of a metre (i.e. 10^{-9}m). So "nanoengineering" means making devices at the size scale of atoms and molecules. The Atomic Force Microscope can image individual molecules and atoms.

[127] When we speak of finding an electron bound to a proton, we no longer today think of the electron as being like a little planet revolving round the proton as the earth revolves around the sun. What we should really say is that there is a high probability of finding a negative charge in a certain region close to the proton, and that the probability of finding it decreases as we look for it further and further away from that region. We can only speak about *probabilities* when talking about atomic events in quantum theory.

Science, Cells And Souls

physicists and chemists,) are identical[128], as are all atoms of carbon (denoted as **C**). So when we look at a hydrogen atom or a carbon atom there is no way to tell whether it is part of a human or not. The properties of "our" hydrogen do not distinguish us in any way from goats or tulips or stones that also have hydrogen in their make-up. The same can be said for carbon and all other elements. Down at this level physics and chemistry throw no light on why we are as we are.

Let's back out a bit further – and here we begin to find something interesting. When we turn our "microscope" onto combinations of atoms, molecules, we reach a stage where there are properties characteristic of the physical make-up of human beings. In order to read reports about biochemistry it is helpful to know how molecules are represented. Pure hydrogen usually exists in the form of a molecule composed of two atoms, written as H_2. There are also many simple small molecules composed of more than one kind of atom, for example *glucose*, which is a combination of six carbon atoms, 6 oxygen atoms, and 12 hydrogen atoms, which we can write as $C_6H_{12}O_6$, and represent by diagrams such as those in Figure 9.1. The bonds shown by the lines are invisible forces, not material links. In these pictures carbon atoms are represented (if at all) by a **C**, and so that the structures in Figure 9.1 are all alternative ways to represent glucose. Subscripted numbers show how many atoms of a particular kind are attached to how many atoms of another kind at that place in the molecule. The exact structure is really three-dimensional and probabilistic, the probabilities being normally so high that we can pretend that the atoms are like little balls stuck onto rods that connect them to other atoms.

It's important to have a feeling for how such diagrams are used to represent chemical compounds, because they appear frequently in biochemistry and biology books. Often carbon atoms are not made

---

[128] "Heavy hydrogen", deuterium, is an isotope of hydrogen with a neutron as well as a proton, but I will not discuss isotopes of hydrogen or other elements. When I say that "All hydrogen atoms are identical" or "all carbon atoms are identical", I refer to their chemical properties, and ignore the differences in isotope number. All atoms of a given isotope number are indistingishable one from another. The points I am making are not affected by this.

explicit in diagrams of complex molecules, as in the second example in Figure 9.1. This figure should help in reading molecular formulae when they appear in other books about biology or biochemistry.

A molecule is the smallest quantity of a substance that shows the typical chemical properties to be expected of that substance. Some molecules exist both in many plants and animals. For example, sugars occur in both plants and animals, and can be isolated from plants such as sugar-cane or sugar-beet as inert minerals. Some sugar molecules are shown in Figure 9.2. Notice that sucrose is a compound of glucose and fructose, linked by an oxygen atom, and produced by removing two hydrogen and one oxygen atoms from the simpler sugars. So we can easily understand that sucrose can be broken down into glucose and fructose, and we can synthesise sucrose from glucose and fructose.

Knowing how molecular structure is represented at this level of analysis we can look in more detail at the structure of chemicals in human bodies. We find molecules such as haemoglobin that show we are dealing with animal tissue: and although we share a vast array of molecules with other species, some *are* specific to humans, and of immense significance, for example the appropriate kind of DNA (through which the genetic programming of the next generation of cells is carried out). To find a molecule of the appropriate DNA in a cell means that we are certainly dealing with a human cell.

When we look at different kinds of material things, we find different kinds of molecules and compounds. A compound is formed by combining different molecules and atoms into a single substance. For example, in looking at rocks we often find silicon, perhaps bound to carbon and oxygen to make fluorspar. In many kinds of plants we find a complicated molecule called *chlorophyll* that gives plants their green colour. We don't find fluorspar or chlorophyll in humans. On the other hand, in many animals we find the molecule *haemoglobin* that makes blood red, and carries oxygen around the body to permit metabolism, but does not occur in plants. A chemical analysis shows that there is a surprisingly close link between the molecular structure of haemoglobin and chlorophyll as shown in Figure 9.3. In this case we change from animal to plant chemistry by changing the metal atom from iron (Fe) to magnesium (Mg).

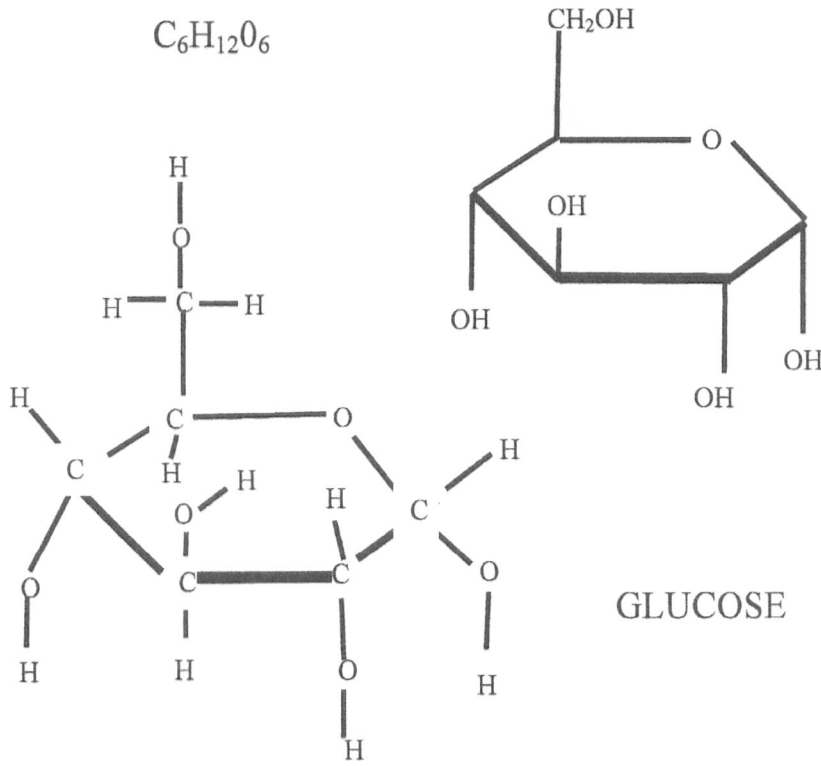

Figure 9.1 Three ways to represent the glucose molecule. Note that in the second version sometimes the presence of **C** and **H** is implicit. The 3-dimensional nature of the molecules is hinted at: for example **H** is above the ring of carbon atoms and **O-H** below it at Position **C**-4 in the glucose molecule, but the reverse at Position **C**-3 next to it.

Clearly the differences between the functions of chlorophyll, which captures sunlight and turns carbon dioxide and water into sugar in plants, and haemoglobin, which transports oxygen from the lungs to other organs in animals, is caused by chemical differences. A Scientific Story is appropriate. Above all, in humans we find many kinds of protein molecules, some of which occur also in other animals, some of which are unique to humans. There are some molecules that are needed for the correct functioning of the human body that our bodies cannot make, and which must be obtained by eating them after they have been made by other species of animals or plants. If we really want to make a person from scratch in the deepest sense, we would have to synthesise all the proteins, lipids, carbohydrates, and so on from their component elements. But of course one way of doing this would be to use ribosomes and other organelles as our machinery for synthesis, and this would not be "cheating", since we know that they are molecular mechanisms, not themselves living organisms.

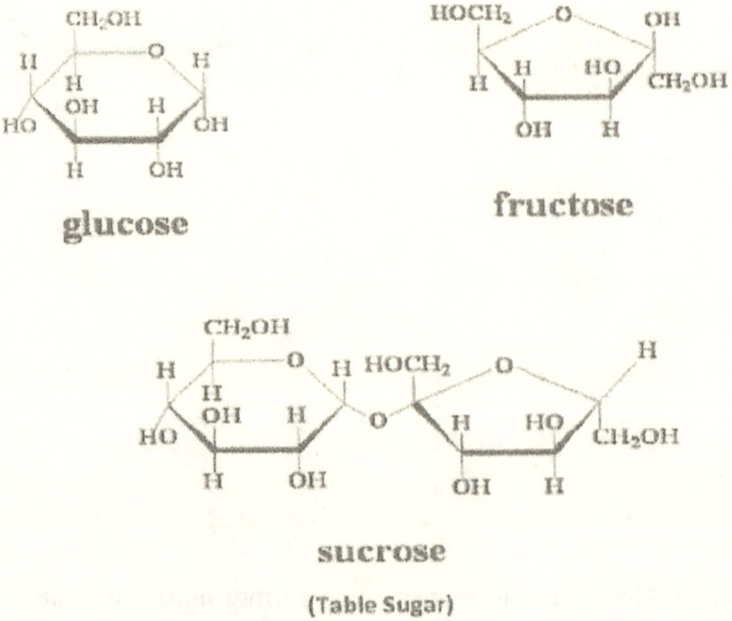

Figure 9.2. The relation of Glucose, Fructose and Sucrose. Sucrose is a compound of glucose and fructose. By combining glucose and fructose and removing a water molecule we obtain sucrose.

Top-down biochemistry

The passage from sub-atomic particles to giant molecules is an unbroken sequence of investigations using physics and chemistry: we don't need any other disciplines or kinds of stories to give a full description of the structure and function of the physical *components* of a person. The investigation is a "bottom-up" sequence, starting from the simplest elemental physical components of matter and ascending to ever more complex structures. If we assembled an appropriate collection of molecules there is no reason to doubt that

Science, Cells And Souls

they would interact as usual in chemical reactions. Now it is time to continue our Scientific Story using a top-down approach, starting from a complete organism.

HAEM   CHLOROPHYLL

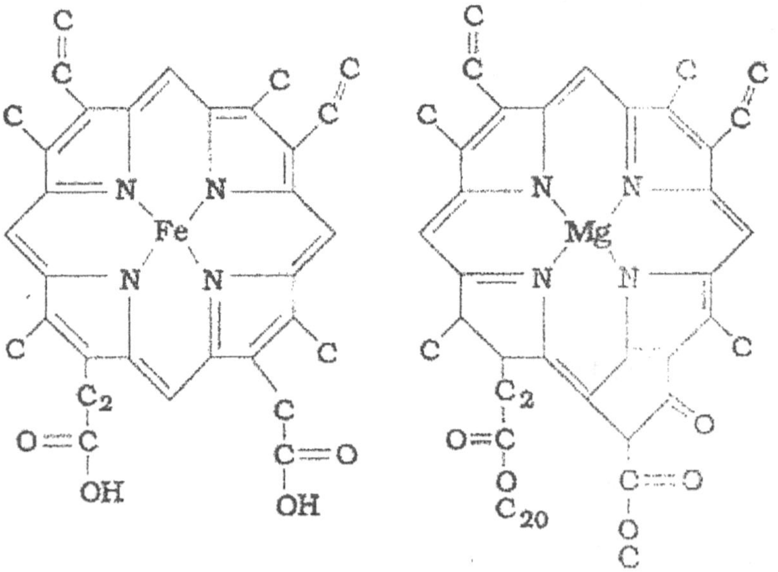

Figure 9.3 Part of Chlorophyll and Haemoglobin molecules. Note the Iron (Fe) atom in haem and the Magnesium (Mg) atom in cholorphyll Chlorophyll occurs in green plants but not in animals. Haemoglobin occurs in the red blood cells of animals but not in plants. Both are variants of the porphyrin molecular structure and they are almost identical except for chlorophyll having a Mg (magnesium) atom in the centre while haem has a Fe (iron) atom.

In a person all the cells of the adult originate from a single cell, the fertilised ovum, the zygote. But during development and by the interaction of genetic prgramming and exposure to different environmental factors in the womb, cells in bodily organs come to differ hugely one from another. For example the white blood cells, *leucocytes*, look and behave almost like *Amoeba*, changing their shape and ingesting bacteria and foreign bodies in the blood. Early biologists called them *phagocytes* ("cell eaters") because of this capacity. On the other hand human red blood cells don't have a nucleus but circulate in vast numbers carrying oxygen to tissues as atoms incorporated into the haemoglobin molecule. Muscle cells have special structures in them that respond to certain chemicals by contracting strongly, which is how muscles generate physical force. (Remember the Story of Saltarella?) Chemicals to stimulate the muscle cells are released at the muscle by the ends of *neurons,* nerve cells that carry to the muscles electrical impulses through a chain of nerve cells that may begin in the brain. The way to make this huge variety of cells would not be to manufacture each kind. Rather, we would make a single zygote and let it develop normally.

For our purposes *neurons* are the most important, because it is through the interaction of nerve cells one with another and with their chemical environment that all the activity of the brain is carried out. The activity of neurons in the brain is the physical vehicle of our mental life, so let's briefly examine their properties.

A neuron (often spelled *neurone*) has a cell body, which contains the nucleus, and a number of short processes which end in very fine branches, visually not unlike the roots of a tree (in fact they are called *dendrites* from the Greek for "tree-like things"). In addition there is one much longer process that is like the trunk of the tree, and this is called the "*axon*"[129]. The axon ends in a spray of fine twigs, at the end of which are small swellings, called "*boutons*" or "*synaptic bulbs*", that are attached to the dendrites or body of another nerve cell in an anatomical arrangement called a *synapse*.

A typical nerve cell receives stimulation from other cells through the boutons attached to their dendrites and, in its turn, stimulates other neurons (or other kinds of cells) by sending a signal along its axon to the boutons at the axon's end. We understand in great detail how this happens. The axon of a typical neuron is surrounded by a sheath of

---

[129] The axon can be very long – the axons of nerve cells in the brain of a giraffe can be several metres long as they run down the spinal cord carrying messages to muscles in the rear legs. Think also of the great whales and dinosaurs such as diplodocus. For a humorous comment, see http://poetry.poetryx.com/poems/8026/ .

fat, the myelin sheath made of Schwann cells, which is interrupted every few millimeters at a Node of Ranvier where the cell membrane of the neuron is bare. When the cell is at rest, there is a difference in the concentration of potassium and sodium ions (positively charged sodium and potassium atoms, Na^+ and K^+) inside and outside the neuron, and this difference is maintained by biochemical activity in the cell. This results in an electrical potential across the cell membrane of a few millivolts. When the cell is stimulated by the arrival of a signal at one of its dendrites the physical properties of the cell membrane change, allowing a massive change in the concentration of Na^+ and K^+, that occurs in a few milliseconds. A pulse of electricity thus runs along the axon, hopping from node to node at a speed of about a hundred metres per second. The electrical impulse arrives at the end of the axon and spreads into the boutons (axon terminals). Here it causes the release of chemicals into the gap between the bouton and the dendrite of the next cell, and in turn that next cell is stimulated[130]. Scientific research has allowed us to construct very detailed models of the physics and mechanics of these processes. We know how the structure of the membrane changes to allow the difference in flow of Na^+ and K^+ ions, what maintains the potential across the membrane, how the neuron is restored to its resting state after firing, and so on. All impulses have the same magnitude: the stronger the stimulus, providing it is above the threshold sensitivity of the neuron, the more frequently the neuron fires, but the impulse is always at the same voltage.

This is of course a simplified account. Some neurons do not stimulate the next cell but rather prevent it from firing: they *inhibit* the next cell by releasing a different kind of chemical at the boutons. Sometimes the stimulation of one cell by another is insufficiently strong to make it fire. There is a period of a few milliseconds after a cell fires when it will not fire again. Some neurons stimulate muscle cells to contract instead of stimulating other neurons; and some cause chemicals such as the hormone adrenalin to be released into the bloodstream. But the important principle is that all the activity of the nervous system, *all the activity of the brain, consists of these kinds of events, namely*

---

[130] You can watch an animation of these mechanisms at http://people.eku.edu/ritchisong/301notes2.htm.

electrochemical interactions among cells. Somehow our entire mental life is carried by this kind of activity, and we understand the physical chemistry in great detail. Whatever the mysteries of mental life may be, they involve well-understood physico-chemical events, and indeed *require* them. If all physico-chemical events stop, the person is dead, *brain dead.* When they cease in some regions of the brain but not in others, the person is unconscious, or paralysed, or unable to speak, or unable to see. So to make a brain we would need somehow to make neurons, or at least components that behave in the same way. Here again the easiest way to do this would be to make a zygote and let it develop autonomously.

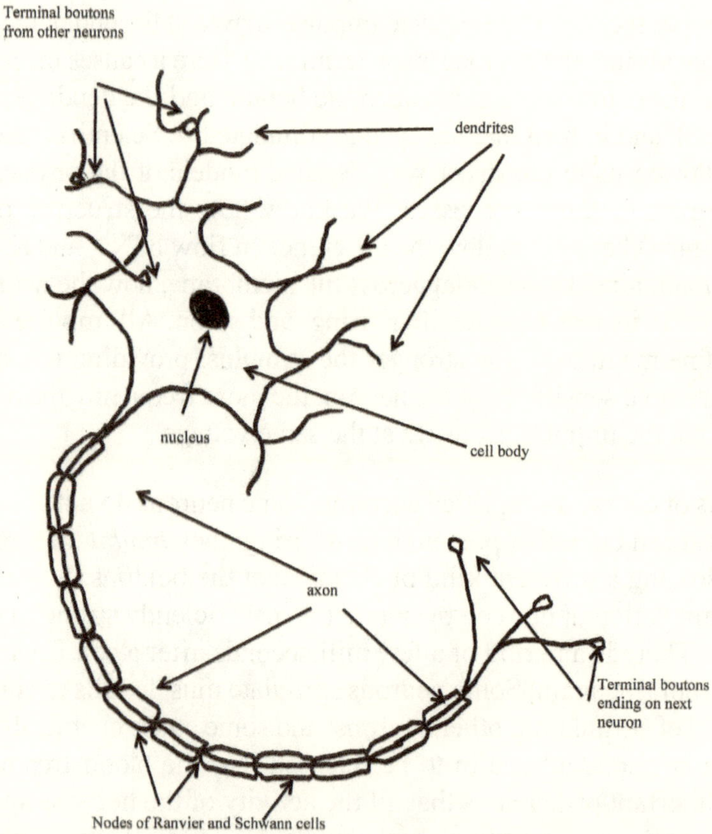

Figure 9.4 Anatomy of typical neuron. The diameter of the body is typically abut 0.01 mm. Schwann cells are electrical insulators. The electrical signal jumps from one Node of Ranvier to the next.

It seems that we can account for the properties (anatomical, chemical, physiological and even behavioural,) of people on the basis of physics and chemistry: we can tell a satisfactory Scientific Story about living humans. Of course there are gaps in our knowledge, but the gaps do not suggest that there we need a different *kind* of explanation. The Scientific Story seems in principle complete or at least completable. For example, whatever the word *mental* means, it does not require us to believe that there are mysterious events, the life of a "ghost in the machine", to account for consciousness, acts of will, and so on. Apparently when certain physicochemical events occur in certain cells we experience mental events and feelings. In describing what people are made of we haven't had any need to invoke immaterial parts or processes: indeed it is difficult to see where we would want to insert such descriptions. *Of course* we have conscious experiences in addition to showing behaviour. *Of course* we have the experience of choosing without feeling compulsion. How to square this with an account of physicochemical events in the body we shall have to discuss later. It is a matter for philosophy, not science. But for now we need to be clear that to be human is to depend on the well-understood laws of physics and chemistry to ensure our brains and body function properly. Our explanations of human nature must start from the facts of physics and chemistry as the basis of biology and psychology. Never forget that whatever your idea of mental life may be, it can be snuffed out by being hit on the head with a brick. That is as clear a proof as one can imagine that physical laws impinge on mental life! A Scientific Story must be the basis of our understanding of human nature.

Building a Person

So could we "build" a person? Could we somehow bring together all the molecular components, all the chemicals, and let them interact; and if so, would the resulting entity be recognisably alive? Well, we would need to make the appropriate chemicals and combine them in appropriate ways. In practice we cannot do this at present, but we can see how in principle we might proceed. In the first place, we would not try to build an adult with millions of cells all assembled artificially. We would not follow the path taken by Mary Shelley's Frankenstein as popularly represented – no bolts through the neck! In Chapter 14 we will discuss how a human develops from a single fertilised cell, and that would be the way to go. If we could make a single living cell, then that cell would do what living cells do, divide, replicate and develop into a more complex organism. If the cell we made was like a fertilized human cell, with suitable DNA, that would be the starting point. There is no reason to think that it is impossible to make any molecule we choose to name, however complex. It would be "just" a matter of chemistry. As we have seen already a DNA molecule has been disassembled and reassembled in an empty cell body to make a new living cell: we know how to design and construct DNA molecules[131]. Indeed in 2014 a new kind of DNA was synthesised that has six base-pairs joining the strands of the double helix, although natural DNA uses only four. The cells into which this new DNA was inserted replicated successfully. We know how to make very simple molecules, such as methane or simple carbohydrates. We know how to make many compounds that typically appear in living cells, such as urea, or sugars, or amino-acids; and there is no reason to doubt that there are appropriate chemical methods that we could use to make proteins, any proteins: after all, that is what bodies do, breaking down food and re-arranging the atoms into new chemicals which are the building blocks of cells and hence bodies[132]. Using radio-active tracers we can follow individual atoms round the body during cellular metabolism. Using an Atomic Force Microscope (AFM) we can watch molecules acting as autonomous biochemical machines. We are developing ever deeper understanding of physico-chemical processes that underlie living processes[133]. For example, Figure 9.5 is a representation of how metabolic processes in mitochondria release energy, using molecules called ADP and ATP, from pyruvate molecules which in turn are derived from carbohydrates. And with the AFM we can see molecular machines carry out the Krebs Cycle.

---

[131] C. Venter. 2013. *Life at the speed of light*. London: Little, Brown.
[132] See animated cartoon versions at http://www.youtube.com/watch?v=-ygpqVr7_xs&feature=related.
[133] P. Hoffmann. 2013. *Life's Ratchet*. New York: Basic Books.

So "making a person" can be described as a physico-chemical process. As Venter has noted, "to make a person from scratch" we don't need to begin with atoms, or even molecules. As I suggested earlier we could use existing organelles. This is not "cheating", any more than it is cheating to use flour and milk rather than carbon and hydrogen when a chef makes a cake "from scratch". In normal life, of course, the way in which the raw materials are brought together to make the basic cell, the gamete or fertilized ovum, is rather more pleasurable and amusing than the work of biochemical engineers. We obtain the raw materials by eating and digesting, and we synthesise the zygote, the fertilised egg, by sexual intercourse.

A final question we might want to ask is, "Where do all the chemical elements come from in the first place?" The answer is that the universe's stock of hydrogen, helium (not used much in our bodies), and some lithium were formed in the first few moments after the Big Bang at the beginning of the universe. All the heavier elements are generated as stars burn hydrogen to heavier elements in thermonuclear reactions which we see as starlight. Explosions of the surfaces of stars, and catastrophic explosions such as supernovae hurl the atoms of heavier elements out into the universe, where they can be captured, perhaps after millions of years, in the gravitational fields of other stars and planets. As Carl Sagan[134] said, "We are made of star-stuff." Given such star-stuff in principle we can use it to make a cell, and hence a person.

---

[134] C. Sagan, 1973. *The Cosmic Connection: an Extraterrestrial Perspective.*

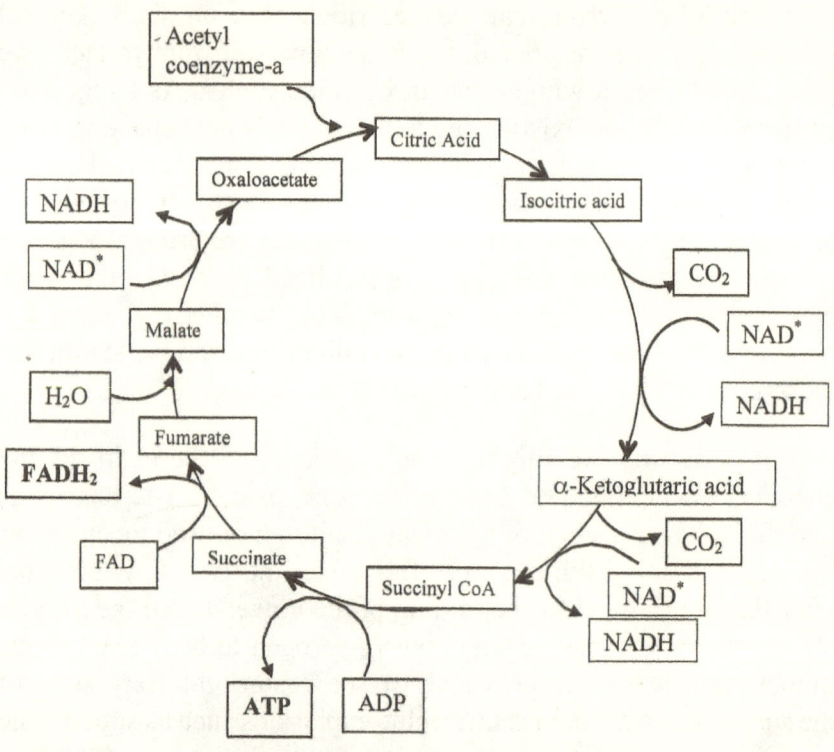

Figure 9.5[135]. The "Krebs Cycle" that provides energy for cellular metabolism. Although complex the chemistry has been completely understood for more than half a century. These processes are carried out autonomously by the biochemical "robots" in the mitochondria[136].

---

[135] Diagram courtesy of Humboldt State University Department of Chemistry. Copyright R.Paselk.
[136] P. Hoffmann. 2013. *Life's Ratchet*. New York: Basic Books.

Chapter 10

Abilities: Nature and Nurture

> L'homme n'est ni bon ni mal, il nait avec des instincts et des aptitudes[137].
>
> Balzac, *La Comedie Humaine.* 1842

So we have a living human. We understand what it means for Saltarella to be alive and how she develops from a single cell into an adult athlete capable of Olympic performance. We can even imagine making a person by making a single human cell and allowing it to develop. But could we endow it with desirable properties?

Nature and Nurture.

An adult human has an astonishing range of abilities. Some like Saltarella's athletic skills are mainly physical abilities closely tied to her anatomy and physiology, but needing her motivation and intention if they are to appear. Others such as those we broadly class as her *intelligence*, one of the Fundamental Words, are mental abilities.

Where do abilities come from? They are not all apparent in the embryo or even the child. In so far as they depend on bodily properties, on muscles, neurons, and other cells, they must be causally closely linked to Saltarella's DNA. That is what we mean when we say that an ability is due to "nature" or "heredity". But we have already seen that strict genetic determinism is false, because of the way that the development of stem cells are affected by their environment: and as soon as we start to think about the biography of an athlete such as Saltarella we can see that genetic determinism isn't true in general of

---

[137] Man is neither good nor bad. He is born with instincts and abilities.

abilities, even physical abilities. Whatever her physical endowments her choice of a coach, her access to role models, her motivation and above all her intense training programme are crucial in developing her ability. We know that the choice of a coach is often critical to athletic success, and that even a week or two's interruption of training can be fatal. "Nurture" is as critical as "Nature".

This contrast been Nature and Nurture, between Heredity and Environment, has been discussed for centuries. What has modern science to say about it? If both are implicated in human ability, can we at least decide which is most important? Let's start with examples of Everyday Stories that raise the questions. "What is more important in determining what a person is like? Is it the family they come from or their education? Their genes or their environment? Biological or social factors?" These questions have often spilled over into politically important questions. "Are some races more intelligent than others?" "Is aggression inevitable in humans?" "Are there genetic differences in the abilities of men and women?" "Can special education compensate for an impoverished upbringing?"

People in Everyday Stories often cite *ad hominem* evidence to support one view or the other. Some are impressed by the way that mathematical ability or musical ability runs in a family over several generations as in the Bach family, and assume that it means that such abilities are genetically determined. Personality traits often seem to run in families, even if they skip a generation. And certainly there was a firm belief on the part of immigration authorities at Ellis Island in the USA in the early part of the 20th century that certain "races" as a whole had lower intelligence than others. All these examples emphasise a belief that heredity rather than environment determines people's abilities.

But equally well others can point to evidence for the primacy of environmental factors. There were periods of many years when tennis was dominated by Australians, but then Americans became dominant, and after that other nations. Middle distance running was dominated by athletes from the British Isles in the late 1950s, but later by athletes from Africa. Today athletes train at high altitudes to maximise their performance. Technological innovation in the 19th

century was centered on Britain, then shifted to the United States, and lately to Asian countries, suggesting that social, economic and educational factors rather than genetics are the more important factors. Women known to have a gene for breast cancer may never develop it; and people with genes for schizophrenia may never show any symptoms if their environment is supportive.

Everyday Stories are not the way to decide on the relative importance of heredity and environment. We need to look for scientific methods to estimate quantitatively the contributions of nature and nurture. But we can see already where the answer will probably lie. As we have already noted, training and coaching can make a huge difference to athletic ability, although the latter is clearly dependent on genetic endowment. Identical twins have identical DNA, but while their genetic endowment may ensure that they both learn language with the same ease, *which* language they learn clearly depends on the country where they grow up. So the idea that we should think of nature and nurture as exclusive opposites in the origin of most abilities is obviously foolish. Abilities are the result of heredity, of environmental factors, and of their interaction in almost all cases.

One can think of the abilities of a person as resembling a construction kit, say a box of Lego®. The genetic endowment of the cell determines the set of shapes and numbers of the various pieces in the box. This sets a fundamental limit on the person's potential. But what is made with those pieces is a matter of how they are used by the environment.[138] Certainly the mother's diet while the baby is growing in the womb is important. The chemical environment in the womb, whether the mother smokes or drinks alcohol and the quality of her diet, are literally vital. The food provided to the growing child, the way the child interacts with her parents and her friends; the nature and style of the schools she attends; even the nature of the social life she leads and the political society in which she grows up will all make a difference to her abilities as an adult. Where a stem cell finds itself determines to some extent how heredity will be manifested.

---

[138] For an excellent critical account of the "nature vs. nurture" debate see S. Pinker. *The Blank Slate*, Penguin Books. 2002.

The whole point of stem cell therapy is that a pluripotent cell, a cell that has the potential to become many different kinds of cells, will become a particular kind of cell depending on the organ into which it is injected, not just on its DNA. At a different level of description a more sociable person when young will have more opportunity to meet different kinds of people and hence have a richer range of experiences than someone more shy or introverted, and so will develop in a different way whatever his or her genetic make-up may be.

In our culture there is something peculiar in our attitude to abilities: we worry so much about inequalities, and whether we can reduce them. We worry about the extent to which heredity or environment may make people ineligible for certain jobs or professions. We feel that we should make allowances for the way in which early education, diet, or culture have been a handicap. But this concern is largely restricted to mental abilities. While many people feel that all should have the same chance of getting to university, or to work in the career or profession of their choice despite disadvantages arising from their upbringing, we don't say the same for their physical abilities. We may feel that "positive discrimination" should provide places at university for those from a relatively impoverished educational environment, but we don't feel that there should be special places reserved on basketball teams for those who are small and undernourished, or for the maladroit in symphony orchestras. Since positive discrimination implies that differences in ability can be offset by environmental ("nurture") factors, why do we feel differently about different kinds of abilities? In particular, why is so much importance placed on *intelligence*.

Intelligence

To look at the nature vs. nurture issue in more detail, let's consider *intelligence*, how to measure it and what we can say about the relative contribution of heredity and environment. What is *intelligence*, and what causes it? It is one of the many abilities typical of humans. Saltarella's skill as a high jumper is an athletic ability, but her ability to think and reason "intelligently" about her plans for her future is a mental ability.

Intelligence is a dangerous noun. Like *life* it seems to be the name of something that is a part, a property of a human. But we know, after discussing *life*, that we should be very careful about talking like that. We never directly see *intelligence:* rather we see different kinds of behaviour that we call intelligent. And what is it that makes us describe behaviour as intelligent? It is the degree to which it supports successful adaptation to the demands of life. Is *intelligence* perhaps just a name for a collection of mental abilities?

An intelligent person exhibits an extensive range of abilities that in everyday life help her to cope with a variety of demands. Indeed we might go so far as to say that the more intelligent a person is the bigger the range of problems she can deal with. That seems to be what we have in mind in Everyday Stories about intelligence. Although we may equate intelligence with success at school or in examinations, we also sometimes say that someone can be very good at academic subjects but at the same time not really very intelligent, perhaps lacking in common sense. It is difficult to define *intelligence* in everyday language. For example the first white explorers who tried to cross the centre of Australia behaved in a way that was extremely unintelligent, rejecting advice and help from the aborigines; and as a result they died. But they could read, write, and do arithmetic. The aborigines could not read and write and do arithmetic, but they were highly adapted to life in the extremely hostile desert. Who were the more intelligent?

If intelligence is the ability to show adaptive behaviour, where does that ability come from? The fact that the Australian aborigines were intelligently able to cope with their very hostile environment while the white explorers were not suggests that the former had learnt specific intelligent behaviour in the context of their environment. On the other hand, later explorers behaved more intelligently which suggests that they already had the capacity for such behaviour, and merely needed to apply it. Intelligence must be related to the way the brain is used to think and make decisions, it must have a biological basis, and hence, one must assume, some aspects of it must be innately coded in DNA and is due to "nature". Can we estimate the relative importance of "nature" and "nurture" in intelligence?

Because of the difficulty of defining intelligence in Everyday language, *all the major research has defined "intelligence" in terms of "IQ",* the *"Intelligence Quotient"*. When we read reports about the relevant importance of nature and nurture we are always looking at research on "IQ". The question of racial differences in intelligence, or how to compensate for early educational deprivation is therefore almost always investigated as differences in IQ using standardised IQ tests. So what is IQ? When you have read what follows you may think that IQ is not the best way to measure "intelligence", but that is a different issue – it is what has been used when people discuss this question, and the only variable on which scientific conclusions have been based.

There are many different IQ tests, of which the most widely used are probably the Wechsler test and the Stanford-Binet test. Typically they have a mixture of verbal questions about vocabulary, grammar, logical reasoning, and verbal understanding, quantitative questions based on arithmetic, and so on. There are some tests such as the Raven Matrices which are only about matching patterns, so as to reduce the importance of language.

Originally researchers calculated a quantitative measure of *intelligence* by defining the IQ of a person as the ratio:

IQ = 100 x (*mental age*)/(*physical age*)

To determine the physical age is obviously straightforward – how old is the person? To obtain the mental age you give tests that estimate the mental ability of the person. In essence all the tests ask how the performance of the person tested compares with the population as a whole. The test score is a measure of the mental age[139].

To define a person's IQ as the ratio of mental to chronological age led to problems in the analysis and interpretation of the statistics

---

[139] You may know of Garrison Keilor's Lake Woebegone, "...where all the women are strong, all the men are good looking, and all the children are above average." Needless to say, that is a satire! So is the observation that since God is infinitely wise and infinitely old his IQ must be 100.

that describe the scores. Hence in more modern work a different definition is used. By definition an IQ of 100 is at the centre (and mean) of the IQ distribution, and therefore we can express a person's score relative to other members of the population simply by how far from the mean his score lies. So in modern work someone's IQ is usually measured simply as how many standard deviations from the mean of the population is his performance.

The way in which IQ scores differ among members of a large population is best described by saying they are *normally distributed*; that is the scores taken over a population fall on a "bell curve" that can be defined by its *mean* value, often indicated by the Greek letter μ, and its *standard deviation*, usually indicated by the Greek letter σ. See Figure 10.1. We introduced these ideas in Chapter 5 when discussing probability, and now we will see their relevance and importance.

Since care is taken to make the test statistically unbiased (more of this later), half the population (50%) have scores above the mean, and half have scores below the mean. About 68% (that is two thirds) have scores that lie between plus or minus 1 standard deviation from the mean, and about 2% have IQs above +2 standard deviations or below -2 standard deviations[140]. The numerical values of IQ that correspond to these values depend on the particular test, but on the Wechsler an IQ greater than 150 is very rare, and so is an IQ less than 50. Because of the way the tests are constructed and standardised there is a big spread: about 60% of the population typically lie between about IQ = 80 and IQ = 120. The tests are constructed so that the mean is at IQ = 100.

---

[140] The formulae for calculating μ and σ are shown in Appendix 1.

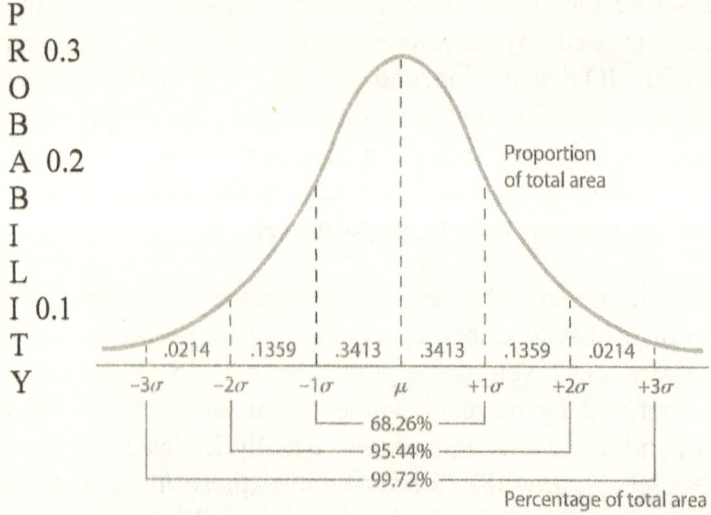

Figure 10.1 The properties of the normal or "bell" curve. The mean is labelled μ: the standard deviation is labeled σ. 68% of the population will have scores between plus and minus 1σ from the mean, and so on (see text).

If as time passes that is no longer true (for example, the vocabulary items in the original test may come to contain words that have dropped from the language) the test is revised and recalibrated to keep the mean at 100 and the s.d. as it was before.

Now consider two races, the *Snacirema* and the *Snacirfa*. The claim that the IQ of the Snacirema as a race is greater than that of the Snacirfa as a race could mean any of the following:

1. The mean IQ of the Snacirema population is higher than the mean IQ of the Snacirfa population.
2. The IQ distribution of one population is very, very different from that of the other: for example, no IQs of the Snacirfa are higher than 100 and no IQ of the Snacirema are lower than 100. This *never* occurs, and if it did one would not accept that the test has been constructed correctly.
3. Although the means are identical in the two populations, the spread of one is less than that of the other. For example no one in the Snacirema have IQs above 120 or below 80, but among the Snacirfa the scores go out all the way from 20 to 180.

4. Although most of the distributions overlap, there is a sufficient difference in the means (or some other statistic of the distributions) to convince us that the difference is too large to have occurred by chance, and so one population on the average has a higher IQ than the other population, both considered as a whole, although the values of the two populations mostly overlap.

The first and last of these claims is the result typically reported by those who believe there are racial differences in IQ. Notice that it could mean that (as shown in Figure 10.2) one population has a mean of 105, and the other a mean of 100, and almost all the scores overlap, but despite the difference being slight they are still statistically highly significantly different (for example there is less than 1 in 1000 chances that this difference arose by chance due to measurement error). Even small differences can be greater than chance if the size of the sample used to calculate the statistics is very large. Here the two distributions almost completely overlap, but it is true that one population, considered as a whole, has a higher IQ than the other. However, if there is certainly a difference, but, say, 95% of the population scores overlap, who cares? It will not make any practical day-to-day difference to the ability or behaviour of a population of, say, 1 000 000 taken as a whole[141].

Test bias.

When we say that tests are unbiased, we mean in a theoretical mathematical sense, namely that the "bell curve" is symmetrical about the mean, and is not systematically skewed towards the high or low end. But there is a more important kind of bias that many think has never been satisfactorily controlled. IQ tests usually use language, vocabulary, arithmetic, etc.. But obviously different cultures make different use of these kinds of skills in their daily lives. If you test a tribe of Amazonian Indians using the Wechsler test of course they

---

[141] If societal progress depends *only* or *mainly* on the very cleverest members of the race, then a small difference like this could make an important difference to the races. But this is not usually true of human populations.

will have a very low score, because they don't do arithmetic and their language does not have any of the words we use, and is not English. They may not even be able to read. Even doing comparative tests on English and French populations is difficult. Attempts have been made from time to time to develop so-called "culture free" or "culture fair" tests. For example, Raven's Matrices use pattern matching of shapes, with no verbal items and no arithmetic. People have to look at pictures of abstract shapes and pick one from a set of alternatives to match a template. I will leave it to the reader to decide whether you think such a test is really a "culture fair" test, and indeed whether there could ever be a truly "culture fair" test across big cultural or "racial" differences.

We also know that practice can often change IQ by at least a few points; and effectively daily life is intensive practice within your own culture. However, if you take someone who has grown up in a culturally impoverished situation such as a city slum, it is not clear that they can be made to respond to coaching in the way that an upper-middle class person in our culture will. That is not because they are stupid, but because IQ tests measure intellectual potential or ability, but do not measure motivation, so are at best only partly a predictor of performance in "real life" tasks. Someone's background may make him largely indifferent to making the required effort, or convinced that the cards are so stacked against him that it is not worth trying. The reason we care about IQ scores is not because IQ is equivalent in all respects to the meaning of the word *intelligence*, not because they really measure *intelligence*, but because IQ scores have been found to correlate highly with other aspects of educational and professional achievements. There is no doubt that if you want to know whether someone will do well at university, or in an intellectualy demanding profession, their IQ score predicts the outcome quite well, but by no means perfectly. If a person chooses to spend his life lying in the sun on a tropical island doing nothing in particular, then whatever his IQ may be, he will not behave in a way that shows high intelligence.[142] IQ at best shows potential. It does not predict or measure motivation. It does not determine life's outcome.

---

[142] Unless of course you think that choosing such a way of life is a sign of high intelligence.

Science, Cells And Souls

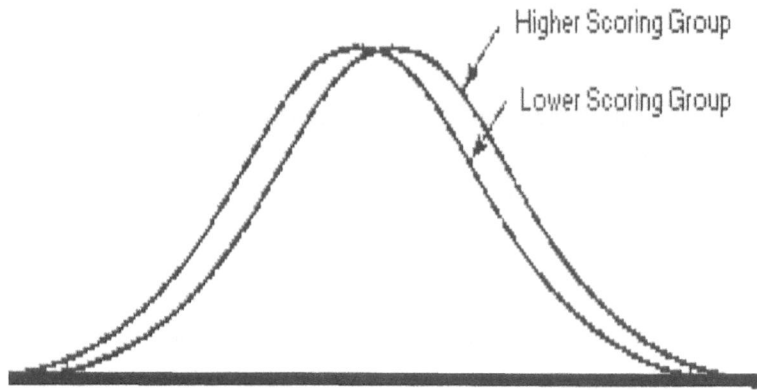

Figure 10.2 The normal curves of two overlapping IQ distributions. The IQ scores are shown along the horizontal axis, and the probability of each score occurring is shown by the height of the curve at that point.

So simple test scores are not sufficient to decide on the implications of small intelligence differences, or even whether the differences are real or have occurred by chance. More sophisticated approaches and different statistics can, however, show differences among individuals, although it is still not clear whether racial differences can be shown, or if they are large enough to make it plausible to say that socio-economic progress can be ascribed to racial differences in "IQ" – and remember that it is IQ, not "intelligence" that we measure, and that as we shall see there are problems with defining "race". At the end of this chapter you will find a fable to make this point clear.

Twins and Heritability

Two kinds of research have shown that there is certainly a strong genetic component to IQ. The first is twin studies. The IQs of identical twins (who have identical genetic makeup, at least with respect to nuclear DNA) tend to be much more similar than the IQs of non-identical twins or siblings. The IQs of identical twins separated at birth and educated separately in different families tend to be very similar. But on the other hand, the IQs of identical twins are not *identical*, exactly equal. And it is certain that IQ scores can be altered by training people to take the tests. There are also some unexplained changes in IQ, such as the *Flynn Effect*, where IQ seems to have increased slowly for the population as a whole over many years. The reason for this is not known, but is very unlikely to be natural selection. So IQ is *not* entirely determined by genetics (or "race" if you believe that what we mean by race is "genetically different"), *nor* is IQ entirely determined by environmental factors (the "blank slate" model[143]).

The second kind of research uses sophisticated mathematical analysis and models of variability (variance) in IQ scores to measure *heritability*. The aim is to estimate what proportion of the variability in scores can be attributed to genetic inheritance, and what proportion to environmental factors. Here too the answer is that some is due to one source, some is due to the other, and some to their interaction. There has been a tremendous amount of disagreement over the analyses, and politically inspired kicking and screaming continues on both sides.[144]

It is important to realise that when someone says that a characteristic such as intelligence is strongly *heritable* this seldom means that the characteristic is unavoidable by the person inheriting it. Indeed the technical term *heritable* cannot directly be used about an individual. If we say that some characteristic, for example *generosity,* has a heritability of 50% in some population this does not mean that half of all the acts of people are forced by their genes to be generous. Nor does it mean that half of the "causes" of a particular action are due to a person's genes, nor that 50% of the population are generous. *Heritability* is a technical term used by geneticists. It means that when we look at a population of people to see how generosity varies among them, half (50%) of the *variation* is accounted for by genetic factors – a statement that is a very long way away from saying that a particular act of generosity by a particular person was caused by his or her genes. As already suggested the way in which genetics affects intelligence in particular and indeed most abilities and characteristics is rather like buying a construction game like Lego®. The bigger

---

[143] S.Pinker. *The Blank Slate*, Penguin Books. 2002.
[144] Again, see S.Pinker, *The Blank Slate*.

the set the greater the range of things you can make. But for any set what is actually made depends at least as much on the imagination and inventiveness of the user as on the range of pieces; and on this analogy heritability is about the range of sizes of sets passed on to the next generation, not what is built from day to day.

The fact that some behaviour is strongly heritable does not directly determine how that behaviour will manifest itself in a particular person on a particular occasion in a particular environment. Many people from families with a gene for schizophrenia will never show symptoms of the disease. Furthermore we may note that the heritability of a characteristic has become built into the genetic makeup by evolution in past environments in the past. It cannot be deterministic for all possible future environments. Saltarella may have inherited the potential to develop a superb athlete's body: she cannot have inherited a compulsion to enter the high jump in the Olympic games, since no evolutionary selection has taken place for that characteristic.

The question of "nature or nurture ?" is actually foolish. If identical twins were *completely* determined by their genes there would be no way to tell them apart except for their physical location – which is clearly not the case. If one of a pair of twins were educated in France and the other in England they would speak different languages, to name but one obvious difference, and that in turn would affect the nuances of their emotional life, how they think, and many other traits. Historically the vicious battle between those who believe in "heredity" and those who believe in "environment" as the main determinant of a person's abilities goes back to the days of Watson and what we can call the Ur-Behaviorists, who said that everything was learned by "conditioning". As one of my colleagues once said, if you believe in complete genetic determinism, you think that the person does not live in a world at all; while if you believe in Ur-behaviorism you believe that at birth the head is empty. Both stances are silly.

Since both genetics and learning have strong effects the only question is how much does each contribute? To that question there is no fixed answer, and indeed it may vary from person to person, from time to time, from environment to environment and from task to task. For example, your DNA may make you more or less sensitive to environmental factors than mine make me. Most of the fuss arose because of political questions as to how best to raise the socio-economic level of those at the bottom of society, particularly in the USA of the 1960s and 1970s. One should always bear in mind that there are no populations where the distribution of scores of the entire population are grossly different from another population if culture fair tests are used. The distributions always greatly overlap as in Figure 10.2.

Variation of Ability: Regression to the Mean

Bell curves describe much about the biographies of living systems. This has an interesting implication when we examine how a person's ability varies over time. We find an effect known as *regression to the mean*. Contrary to what common sense might lead one to expect, if someone one day does exceptionally well at a task, almost certainly he will do less well on the next day. Likewise, if he performs unusually badly on one trial, the most probable result on the next trial will be an improvement.

The reason for this is nothing to do with the amount of effort a person puts into the task, nor how motivated he is. Hold all such factors constant and the same result will be found. The reason is purely statistical and applies equally well to non-biological systems. Suppose we measure the height Saltarella jumps. Her many jumps will tend to form a normal distribution. An unusually high jump corresponds to a point well above the mean, perhaps even $+2\sigma$. Now consider the next jump. Since there is far more of the distribution below $+2\sigma$ than above it, it is far more probable that the next jump will be lower than the previous one. This is purely a matter of statistics, not of how hard she tries. The only case where one should expect an exceptionally high jump to be followed by an even higher one is when there is a strong practice effect, so that very rapid learning is occurring. In such a case the entire bell curve drifts upward. But trial by trial there will be regression to the current mean. This is important in all kinds of training and motivation, and is directly relevant to discussions of reward and punishment.

Race and intelligence

The relation between race and intelligence has been discussed in its modern form for some 60 years. There was earlier research, but modern work uses techniques that were only fully developed around the end of the 1950s. Since then there have been at least three major disputes about the evidence, at intervals of about 20 years. The research has been intense, and so has political interference with the science. People have lost their jobs because of the work they have published. Attitudes have been fairly disgraceful on both sides of the question[145].

The outcome has been in every case uncertain. Some people have concluded that there is no doubt that intelligence is mainly determined genetically, and some have concluded that intelligence is mainly determined environmentally (by education and social factors). The overall conclusion can really only be represented by a statement such as: *there is little or no doubt that there is a strong genetic component to intelligence measured as IQ, but equally there is no doubt that there is an environmental component; and over all there is no reason to think that the differences between races are sufficient to account for example for the fact that "Africa" has not followed "Europe", "Asia" and "America" in their "successful" development of "capitalism".*

I put several words in quotes in the above paragraph because I do not know what *you* mean by the words. *All* of Africa without exception? *All* of Europe without exception? *Your* brand of capitalism? And so on. Indeed the entire research issue has always been bedevilled by problems of definition.

---

[145] I have even heard someone say that were there evidence that black people had lower IQs than white people the results should be suppressed, because it was not politically acceptable to say so, even if true, and even if remedial education could reduce the difference.

Consider *race*. In effect, *race* in this context has always meant population groups such as Black Afro-Americans, White Caucasian Americans, Japanese, Irish, Indians, Chinese and Australian Aborigines. Until very recently these categories were acceptable classifications, for example as used in the census, and the underlying issues concerned such things as whether the socio-economic condition of Black Americans could be brought up to that of Whites by intensive education, or whether there is a genetic lack of "intelligence" that cannot be overcome by educational intervention. It is only very recently that the biologists have had an impact by suggesting that there is perhaps genetically no such thing as "race". The difference in DNA between what have been called different races in Everyday Stories are very small, and when unbiased tests are used to make comparisons the results resemble Figure 10.2. In Everyday Stories "race" still means the difference between "Africans", "Arabs", "Japanese", "English" and so on; and whatever the underlying biological structure may be, these peoples certainly seem to be different, fairly homogeneous population groups if viewed superficially in Everyday Stories. So that is what research has considered. Research has found more difference, on the average, in DNA among members of one of these "races" than there is between the average of the members of two 'races'. However very recent work suggests that there may be some small stable genetic differences due to interbreeding between *Homo sapiens* and different species of Neanderthals some 10 000 years ago. The notion of race in the human species is ill-defined.

Perhaps we can summarise what we have learnt in a somewhat light-hearted way with a quotation and a fable. The quotation is something that was once said by Kolers, a psycholinguist: "People who believe most strongly in the genetic determination of intelligence tend to be those with unusually intelligent children."

The second is a look at ourselves:

Race and the Roman Empire: a fable.

Julius Caesar was talking to Brutus about his campaigns in Gaul and Britain. "Tell me, Brutus," he said, "Why do you think that the white races in the North are so backward? They don't seem able to benefit from our colonisation. Down here you have the Mediterranean races, the Egyptians, the Greeks, the Carthaginians, and ourselves, all with beautiful cities, art, literature, drama, good roads, efficient armies and navies, and lots and lots of money. No one has ever been as rich as we are. But we build houses and roads in Britain, and in Gaul, and nothing happens. They just don't learn! I think there must be something in their genes that means that Northern races can't benefit from civilisation. There seems to be a real difference between the pale northern races and the olive-skinned Mediterranean races that prevents white northerners from learning how to behave in civilised ways."

Chapter 11

Artificial Intelligence: Sharing the World With Artefacts

> ....What I ha' seen since ocean-steam began
> Leaves me na doot for the machine - but what about the man?
>
> Rudyard Kipling. *McAndrew's Hymn*. 1893.

> Like a dog walking on its hind legs it is not done well, but that it is done at all is remarkable.
>
> Dr. Samuel Johnson. Boswell's *Life of Johnson*.

Prosthetics and Robotics

There is another way to throw light on human nature, and that is to ask how nearly we can copy or replace human abilities and characteristics by artificial means. Can we build machines that perform not only the bodily but also the mental and spiritual functions of humans? We have always shared the world with other kinds of animals, but today we find ourselves sharing it increasingly with artefacts[146] that mimic, replace, or supplement biological systems. We are used to Long John Silver's wooden leg in *Treasure Island* or Captain Hook's artificial hand in *Peter Pan*. We are even not too surprised when modern engineering allows para-olympians to run almost as fast as the best Olympic sprinters on artificial "blades" which replace their lower

---

[146] Some people spell the word *artifact*, but I prefer to emphasise the skill needed to make such things, hence *artefact*, to underline the artfulness of the enterprise. (From the Latin, *ars, artis;* hence *arte-fact*, "made by artful means".)

Science, Cells And Souls

legs, using new materials such as carbon fibre. Artificial heart valves, and indeed artificial hearts are well known, while artificial hips and knees are commonplace. But today we see the first wave of a different kind of artificial device, one that emulates what we think of as mental functions. Already primitive vision has been restored to a blind man by burying a photosensitive electronic chip in his retina[147], artificial hands controlled by the owner's thought are becoming increasingly common [148], speech recognition is available on home computers, and computer programs have beaten chess Grand Masters.

Artificial limbs and body parts are examples of *prostheses,* artificial devices that replace natural parts of the body. (Artificial devices that supplement existing bodily functions are properly called *orthoses.* Stilts are an ancient example of an orthosis, and exoskeletons, frameworks resembling suits of armour with powered joints that can be worn by users to greatly supplement their physical strength are a modern example[149].) Closely related to such devices are machines that are autonomous, not controlled by a human user, and which we call *robots,* a word introduced in 1920 by the novelist and playwright Karel Capek[150]. Robots today include many kinds of devices from autonomous spacecraft to humanoid servants and pets[151], various military devices, the humble automatic dishwasher or clothes washer, and recently developed autonomous vehicles capable of driving on roads in traffic[152]. In late 2013 a robotic aircraft flew to an aircraft carrier and landed itself with no guidance from a human. For brevity I will use the word *robot* to refer to any hardware device designed to support, replace or mimic human ability and that does not need moment to moment input from a human controller.

---

[147] http://www.bbc.co.uk/news/health-17936302.
[148] http://www.nlm.nih.gov/medlineplus/news/fullstory_132301.html,
[149] http://gizmodo.com/5587600/rex-bionics-has-the-technology
[150] Capek, K. 1920. R.U.R. (Rossum's Universal Robots).
[151] http://www.technologyreview.com/news/522086/an-artificial-hand-with-real-feelings/
[152] http://www.sciencedaily.com/releases/2011/05/110511111957.html

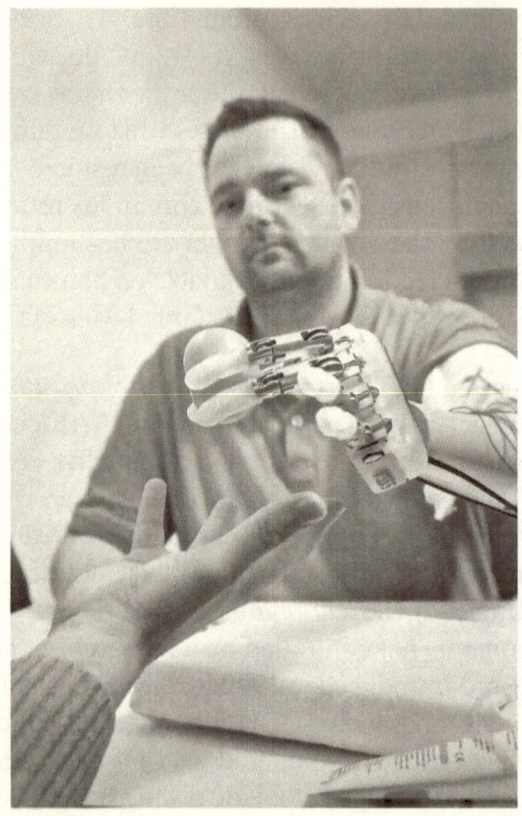

Figure 11.1 A prosthetic hand activated by muscular tension expressing the intentions of the user. Signals from sensors in the artificial fingers send tactile feedback into the sensory nerves of the arm so the user can feel the orange and so modify his grasp[153].

We must also think about systems that mimic or supplement behaviour and function but not the structure associated with human function. This is often called *artificial intelligence*[154], or *AI*. The whole field of artificial simulation or mimicry of human and other biological systems is sometimes called *cybernetics,* after the title of a book by Norbert Wiener, although this usage is strictly incorrect. (Wiener

---

[153] Picture provided by Ecole Polytechnique Federale de Lausanne, 01_Copyright_LifeHand2.jpg. Videos are available from that source.
[154] I have elsewhere called AI *applied logic.*

originally defined[155] cybernetics as *the science of communication and control in the animal and the machine.)* AI usually refers to the logical simulation of human ability rather than the building of a hardware artefact to perform tasks, and the usual way to do this is by writing a computer program to display the ability. A computer program that plays chess is an example of AI.

Questions about AI and robotics fall naturally into three classes. First, are there limits on developing prosthetics? In particular is there any reason why we could not replace parts of the brain with artificial hardware components? Second, what are the limits on AI? Can we make a computer or other machine perceive, think, or choose? Finally, given that cybernetics and robotics are only a few decades old, what can we expect in the future? Will robots in the future surpass the abilities of living systems, particularly those of humans? For getting insights into human nature robotics and prostheses are perhaps of little interest, except for the question of whether we could replace parts of the central nervous system.

One of the most extreme views about the possible development of AI and robotics is that of Kurtzweil[156]. He believes that we are about to reach a time when the abilities of machines, particularly AI, will exceed the abilities of humans, and that when that happens a new era will arrive, in which humans will perforce have to allow themselves, in many situations, to be controlled by hyper-intelligent and probably sentient machines. By encoding their knowledge, desires, and abilities in such machines, humans will even achieve a kind of immortality. Interested readers should evaluate Kurtzweil's claims for themselves.[157] But always remember that one should not exclude possibilities because of *a priori* assumptions. What we need is evidence and analysis, not prejudice or startled disbelief. Here I want to see what we can conclude about human nature from the characteristics of AI, prosthetics and robotics.

---

[155] N. Wiener, 1958. *Cybernetics: communication and control in the animal and the machine.* Cambridge, Mass. MIT Press.
[156] R.Kurtzweil. 2011. *The singularity is near.* New York. Penguin Books
[157] A web site is http://bigthink.com/videos/ray-kurzweil-your-brain-in-thcloud

Can a human biography be simulated on a machine, at least in principle? What would make such a study particularly interesting is that it would allow us to discuss topics that are usually confounded by Everyday Stories and their assumptions. For example, if we ask questions about the nature of thinking we may assume that only living things can have a mind, so only living beings can think. After all, Descartes concluded that animals were just unthinking and unfeeling machines. It may be helpful to think about mental actions without making any such assumptions. What then would count as "thinking" by a machine? And what would it imply about the nature of human thinking?

Let's go back briefly to the earlier chapters in which we discussed the way that language may bias our discussion. Remember the importance of definitions. If we want to ask whether the notion of artificial intelligence, AI, makes sense even as just an idea, we need to define what we mean both by *artificial* and *intelligence*. If we try to decide whether a machine can think, see or feel, we must define *think, see* and *feel*. And we should not be side-tracked by thinking in terms of literally copying human functions in all their details. We do not deny that aircraft fly just because they don't flap their wings; so we should not *necessarily* be concerned with the fact that machines do not have brains when we ask whether they can think or show abilities that are normally associated only with humans or other living creatures. More precisely we need to define what counts as doing the same thing or having the same characteristics. We would not normally think of a camera as "seeing" (in the sense of "being aware of") the picture that it records, but we would probably not hesitate to say that a tiger really "sees" its prey. If a robot responds to visual information by appropriate behaviour, should we think of it as a camera or a tiger?

Let's start by thinking about replacing parts of the body with artificial components. There is nothing strange about the idea of an artificial hand to replace a real but damaged one; indeed such a prosthesis would be welcome to many people. Artificial hips and knees are taken for granted. Again, we should not insist on the exact anatomy being replicated. The tentacles of an octopus have sensors and suckers along their length, and an efficient artificial tentacle might

Science, Cells And Souls

have advantages over a standard prosthetic arm that mimics a human arm and its joints. Probably the major disadvantage would be how un-natural, indeed monstrous, it would look[158]. But what would we think about artificial replacements for parts of the nervous system? We regularly embed microchips in animal tissues to identify pet dogs, but not into their nervous systems. What would we expect to happen if we put an electronic chip into the brain? What would be the function that we would be replacing? If we replace a nerve cell do we replace a thought or a memory, or merely a mechanism?

Let's think again about a typical nerve cell, such as that described in Chapter 9. A neuron is made from biological molecules, has a cell body and dendrites that are specialised to receive electrical or chemical stimulation, and when stimulated transmits an electrical impulse to the far end of the axon, the long process that connects the cell body to the body and dendrites of another neuron, to a muscle cell, or to cells that release chemicals, hormones, into the blood stream. The initial stimulation may be caused by chemicals released from the end of the axon of a prior neuron, or may be due to energy received from the external world by receptors in sense organs, such as the rods and cones in the retina that transform the electromagnetic energy of photons into nerve impulses, or the hair cells in the ear that transform mechanical energy from sound waves into auditory nerve impulses. The final output is always to release some kind of chemical to stimulate another cell. Overall then the function of a neuron is to receive stimulation and in turn stimulate another cell or cells. Can we imagine replacing nerve cells with neuronal prostheses?

Not merely can we imagine so doing: it has been done. We have been able for some years to put tiny prostheses, called *cochlear implants*[159], into the ear to cure some kinds of deafness. And recently a man who was completely blind because of retinal abnormalities has had a photosensitive electronic chip resembling those used in digital cameras placed in his eye behind the retina and connected to

---

[158] Something of what this would be like can be seen in Japanese erotic art involving coupling between humans and octopodes.

[159] http://en.wikipedia.org/wiki/Cochlear_implant

the latter and to the optic nerve[160]. He can now experience coloured lights, although he does not yet have clear detailed visual pattern perception. On the output side of the nervous system there have been many demonstrations that nerves can be coupled to electromechanical prostheses so that a person can control an artificial limb by thinking about required actions. Figure 11.1 showed a system which gives the user feeling by passing electrical signals from artificial sensors into the sensory nerves of his arm, while he controls motors that drive the motions of the fingers by flexing his muscles. Without looking he can tell whether objects are soft, malleable, hard, etc. nine years after losing his hand.

Perhaps on reflection none of this is particularly surprising. If we can couple an electronic device to a nerve cell in such a way as to stimulate it that neuron will transmit an electric pulse along its axon to its terminal boutons. If we were able to insert a tiny electronic device that detected electrical signals from the end of a nerve axon and transmitted them along a micro-wire which ended on the dendrites of another nerve cell, we could certainly make such a micro-prosthesis take the place of a biological neuron to some extent, although it would be difficult to make a wire release chemicals to stimulate the dendrites. It would probably be necessary to stimulate the next cell electrically, although this would not prevent the prosthesis from working. There would be immense practical difficulties, since a neuron usually receives inputs from hundreds of neurons, and may be, in its turn, connected to hundreds of others that it stimulates. It would probably be impossible ever to match all the connections even for a single neuron. But since the nervous system is particularly efficient at adapting itself, there is every reason to think that replacing a living neuron with an artificial one would allow the brain to continue to function. The technology of developing devices at this scale is a growing industry, called *nanoengineering*.

In principle we should be able to replace receptors in sense organs with artificial sensors, and neurons in the brain's networks with artificial neurons. What we will do is to replace one physical system, made

---

[160] http://www.bbc.co.uk/news/health-17936302.

of proteins and other biological molecules with another composed of silicon, plastic, germanium, etc.. Each of them transmits an electrical signal from one end to another. There is nothing magical or mysterious there, and so our neural prosthesis should let the nervous system continue to function. So let's replace another one. . . And then another one. . . And then. . .

Where, if anywhere, would this process have to end?

Let's take a different point of view. The brain consists of an immensely complex interconnected network of some 10 000 000 000 nerve cells. The bioelectrical interactions of the cells in this net are the vehicles of our mental life. Is there any reason to think that we could not make from electronic neurons artificial neural nets that could do what biological ones can do, for example recognise patterns such as geometric shapes in vision, or tunes in hearing? Here again we already see such abilities in Everyday devices. Some digital cameras can recognise that they are being pointed at a face, and even whether the face is smiling; voice recognition is being used in security systems and telecommunication systems; and speech recognition is available as a way to dictate commands and text to computers and word-processors and to interact with telephone users. Robots can learn to find their way around natural and built environments avoiding obstacles even when the latter are unexpectedly encountered, and as mentioned above a robot aircraft recently landed itself on an aircraft carrier.

There has been a huge amount of research into the abilities of artificial neural nets in recent decades, although these nets are usually not built with physical artificial neurons, but are computer simulations. We can build artificial systems that mimic much human behaviour that in the past was thought of as connected with "mental" functioning, such as pattern recognition and some kinds of thinking, reasoning and problem solving. Nerve nets have turned out to be extremely powerful learning systems that adapt themselves to the inputs they receive and learn to recognise patterns without human intervention. Indeed when an adaptive nerve net simulation has finished learning, often the programmers no longer know exactly what is happening in the simulated nerve nets. The nets show self-organization.

What can we expect the future to hold? On the one hand we can expect technological advances to let us build hardware that we can insert into existing nervous systems to replace biological components, and yet still allow the brain to function. If you find it hard to envisage putting microchips into the nervous system, think instead about possibly building microchips not from silicon, but from protein and other biochemical components. Research into biological computing is already under way. Such chips would be just as artificial as silicon microchips, because, as we saw in Chapter 7, there is nothing special or magical about biochemistry: it is just the manipulation of natural physical substances. In principle we can connect any carbon atom to any hydrogen atom. But the second way of approaching the question of the limits of what has been called "the sciences of the artificial"[161] is to concentrate less on the physical engineering and more on the conceptual and functional issues. Is there any limit, in principle, to what we can make a machine do[162]? Can we make a machine think? Does a camera "see" the smile on a face, or merely respond to it?

Turing Machines

Many people's first response to such a question is that it must be impossible to answer, because we don't know what kind of machines may be built in the future. But for a certain class of machines we can bypass that problem. Most research into AI is carried out using digital computers, and for digital computers we can say precisely what the limit will be, even in the future. This is due to the work of Alan Turing, and digital computers are known more generally as examples of *Turing Machines*.

For several decades the work of Turing was known mainly to mathematicians and computer scientists, but recently his name has become deservedly well known to a much wider audience[163]. In

---

[161] H.A.Simon *The Sciences of the Artificial*, MIT Press, 1996. W.S.McCulloch. 1965. Embodiments of Mind. MIT Press.
[162] An introduction to this topic has recently been published by N. Bostrom. *Superintelligence: Paths, Dangers, Strategies*. Oxford University Press. 2014.
[163] A. Hodges, *Turing, the Enigma*. Princeton University Press. 2012.

addition to his work on cryptanalysis during World War II, he made important contributions to the foundations of mathematics, and in the course of that work he discussed the fundamental nature and limits of computation. In particular he described what came to be called the "Universal Turing Machine". He proved that a computational system could exist that could compute anything any other conceivable computational system could compute, even if the latter had not as yet been invented and might be much bigger and more complex. A Turing Machine is a device that manipulates symbols, reading them from a tape, writing them to a tape, and moving the tape from one position to another. It is a theoretical device, and no one would build one for practical purposes; but it is in fact an engine for performing logic, and can in principle solve any problem that can be solved purely by using formal logic[164]. We would nowadays say that it manipulates "bits", as do our computers. Now it can be shown that digital computers are Turing Machines, so that for any problem in AI, and for a wide variety of other philosophical questions, we can ask whether a Turing Machine could perform the calculations needed to solve the problem. If so, then in principle we could build some machine, either a real physical computer or a machine simulated on a computer that would be able to perform that function. We can tell whether in principle we can build a specified prosthesis even without knowing what kind of machinery may be invented in the future. If we want to know whether we will one day be able to build a machine to perceive, think, choose, or create or appreciate works of art, we can find the answer by asking whether a Turing Machine could do so. And that calls for a philosophical analysis of the problem.

To better understand this it may help if we see how artificial neurons can be regarded as devices equivalent to Turing Machines. "Pitts-McCulloch neurons"[165], (PMNs), were invented by a logician Pitts

---

[164] Formal logic can be thought of as a kind of algebra or calculus using operations such as AND, NOT, OR, NOR etc. operating on abstract symbols. Properties such as TRUE and FALSE can be represented and calculated using the calculus. For example, if it is False that a statement is False, then the statement is True.

[165] W. McCulloch, 1967. *Why the mind is in the head.* In L. Jeffress, ed., *Cerebral Mechanism in Behavior.* The Hixon Symposium. New York. Hafner.

and a cyberneticist McCulloch who described the nervous system in terms of the computations it must be carrying out. PMNs were not embodied as hardware, but could be simulated in a digital computer. These artificial neurons are equivalent for many purposes to Turing Machines, and we can summarise the implications of their work and that of Turing in the following three propositions.

1. Any Turing Machine can perform in principle any logically possible computation, regardless of how powerful a machine is required to perform it.
2. A suitable network of Pitts-McCulloch neurons is a Turing Machine.
3. A suitable network of Pitts-McCulloch neurons can be described to show any behaviour that can be defined in a finite number of words.

So we can imagine constructing neural prostheses out of artificial neurons that resemble PMNs, and then we need to look at the relation between real neurons and PMNs. In so far as they are equivalent, we can ask some completely general questions about the relation of AI and robotics to human nature by considering PMNs and digital computers. Pitts and McCulloch concentrated on the fact that however the biochemical mechanisms in a neuron operate, what each cell does is to receive one or more inputs as a stimulus, and respond to the latter by producing an output. That output can either stimulate another neuron to produce an output in its turn, or inhibit another neuron, preventing it from firing. A PMN is a simple digital device, and the authors described several kinds of specialized PMNs. To show how they can be used to simulate computation, we will look at a so-called "AND gate". This is a PMN with two inputs, and if they are simultaneously both electrically positive, the single output in turn goes positive. If one or both are at zero or negative volts, then there is no output. The output can be used as an input to another cell. Another kind of PMN has an output that prevents neurons to which it is connected from firing, so behaving as an inhibitory cell.

Figure 11.2 shows how we might imagine such a PMN, embodied as a microchip.

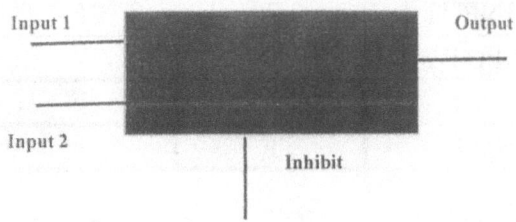

Figure 11.2. Schematic Pitts-McCulloch neuron.

We can write a sentence that describes the calculation being performed:

> "IF Input 1 is HIGH
> AND Input 2 is HIGH,
> THEN set the Output to HIGH,
> IF and ONLY IF the input to INHIBIT is
> NOT HIGH."

Such a sentence is in fact a computer program and so PMNs are logic computing machines. The similiarity in function to a biological system is obvious. Inputs 1 and 2 might come from neurons from adjacent spots on the retina, carrying visual information. The Inhibit input might be from the auditory system so implementing a kind of attention mechanism, inhibiting vision when there is an auditory input.

We are today familiar with the idea that digital computers calculate using binary arithmetic, and manipulate "bits" of information. The following Figures show how input/output relations, voltages, logic and binary arithmetic are equivalent in PMNs. In each case the entries in the cells of the matrix show the Output for the combination of Inputs 1 and 2 outside the boxes. T stands for TRUE and F for FALSE.

INPUT 1	T	F
INPUT 2		
T	T	F
F	F	F

Table 11.1 Truth table for an AND function. The body of the table is the output. The output value of the function is only TRUE if both inputs 1 AND 2 are TRUE, otherwise the value is FALSE.

INPUT 1	+5v	0v
INPUT 2		
+5v	+5v	0v
0v	0v	0v

Table 11.2 Truth Table of a Pitts-McCulloch neuron expressed in terms of input and output voltages

INPUT 1	1	0
INPUT 2		
1	1	0
0	0	0

Table 11.3 The Truth Table of an AND function expressed as multiplication in binary arithmetic.

If we re-write TRUE and FALSE as 1 and 0, we get the third table, and we see that logical computation is equivalent to binary arithmetic. If we restrict ourselves to using just the symbols 0 and 1 rather than the letters of the alphabet we can still write any message that can be written in ordinary language, since we can replace the letters A to Z with the numbers 1 to 26, write the latter as binary numbers, and express everything with just the two symbols 0 and 1.[166] We have assumed in these examples that there is a zero input to the Inhibit line in all cases. If there is a T, +5v or 1 input to the Inhibit line, then the output is always F, 0v or 0 respectively.

Since digital computers operate using binary arithmetic, we could simulate PMNs in a program, and in turn use the program to simulate the function of real neurons that can receive inputs and respond by having their outputs stimulated or inhibited. Using such simulations we can define functions equivalent to memorizing, conditional choice, learning, reasoning, pattern recognition and other operations typical of intellectual functioning.

Although all the subtleties of variable voltages, multiple connections, biochemical operations, etc., found in real neurons are missing, they can be simulated to any degree of accuracy required in a computer program. Of course if we think not in terms of biological function but in terms of the logical functions being computed, then we avoid having to think about biology and concentrate on logic. And we can use the latter to explore what in principle can be solved by AI, even examining what will be possible when we have more powerful machines. In practice we ignore the details of Pitts-McCulloch neurons and simply write computer programs that embody logic[167]. But a program is always equivalent to a Turing machine that could be constructed if required. So let's start by asking whether such machines can show some typically human abilities. It seems reasonable to think that they could show any logical abilities, and already some human logical abilities have been surpassed by computers[168]. But is logic

---

[166] This is of course how digital computers work.
[167] Strictly speaking we do not need to discuss the abilities of computers via Pitts-McCulloch neurons. I have done so here because some readers may find it easier to approach the idea of AI if they think of it being performed by artificial neurons rather than merely by programs. The two approaches are identical in their implications. There are a variety of logics known today such as multivalued logic, fuzzy logic, etc., which we cannot study in the present book. Their existence has no implications for the arguments presented here.
[168] N. Bostrom. *Superintelligence: Paths, Dangers, Strategies.* Oxford University Press. 2014.

the whole of human mental life? In Chapter 3 we saw that some supposedly mental functions, such as a key's ability to open a lock, or human bravery, might not be material things, but at the same time might not be non-physical. How would these fit into a Turing Machine framework? Can *bravery* be described as a logical function?

Artifical Intelligence

It is now common to encounter machines that can recognise patterns. Some cameras can tell whether or not there is a face in the direction they're pointed, and even whether or not the face is smiling; and speech recognition software is becoming increasingly efficient, turning spoken speech into written text or commands to computers. Today there are many examples of efficient pattern recognition software and hardware in commercial, scientific and military applications. Humanoid robots make extensive use of artificial pattern recognition. Almost the first stated intention of computer scientists once computers became available was to create intelligent problem solving behaviour, and among the earliest examples were programs to play chess, prove theorems in logic, and solve complex problems such as medical diagnosis. Computer scientists wanted to make machines that could think[169].

Did they succeed, and were those computers intelligent? Well, certainly the best programs produced some output such that if we found it in the desk of a human, written out by hand, we would have said it was evidence of intelligent behaviour. Indeed in several cases even the early attempts in the 1950s and 1960s were impressive, and showed that a common criticism, that computers "can only do what they are programmed to do" and cannot show inventiveness and originality, is false. One of the first attempts at AI was a program to prove theorems in mathematical logic, and it not merely succeeded, but in one case found a new proof for a theorem by Russell and Whitehead in *Principia Mathematica*; a proof that had not been known until then, and that was judged by experts to be more elegant

---

[169] E.A.Feigenbaum and J.Feldman. *Computers and Thought.* McGraw Hill, 1963.

than the original[170]. This brings us to what is now known as *Turing's Test*, a criterion for whether machines can think that was proposed by Turing himself.

The Turing Test

Suppose that we ask someone to solve a problem that requires thinking. We are shown the work he or she does and the solution proposed, but we never see the problem solver, only the work produced. We do not doubt after looking at the work that thought was required to solve the problem. Suppose that we find such a paper on which is typed out an argument in logic or mathematics, and on reading it decide that it is evidence of deep thought. We might feel even that it would be interesting to meet the person who was capable of such intellectual skill. "She must have been a pretty clear thinker to have found that proof." Perhaps we would even feel the same about someone who solved a relatively simple riddle, such as the old one about how to cross a river in a boat that can only carry the rower and one passenger, when what has to be taken across is a fox, a goose, and a bag of corn[171]. Most people, particularly if they fail to solve the riddle, would say it needs thought. And at least to begin with, even to play Noughts and Crosses [172] requires thought. (The only alternative is to guess.)

Now suppose that, having seen the solutions written down, you are told that the solutions were found not by a human, but by an AI problem-solving computer program. Since you have already agreed that thought is required to find the solutions, should you not agree that the computer was able to think? If something behaves as if it is thinking, and if there is no other obvious way to solve the problem, we may as well accept that thought has occurred, even though the thinker

---

[170] en.wikipedia.org/wiki/Logic_Theorist
[171] If left alone the fox would eat the goose and the goose would eat the corn. Just in case you have forgotten the answer, you take the goose across, and return with an empty boat. You take the bag of corn across and bring the goose back and leave it on the bank. You take the fox across and return with an empty boat. You take the goose across.
[172] Tic-Tac-Toe

was not alive. The alternative is to define thought as a function only to be ascribed to living creatures. In that case AI will always be, by definition, "as if" thinking, not genuine. However to take this step should not be arbitrary, but the result of careful analysis and justification. The Turing Test asserts that if you cannot tell whether the machine or the human performed the task, then you may as well say that if the machine did it then it has done what the human would have done had the latter done it. Or perhaps you might want to say not that the machine had been thinking, but nonetheless that it had shown intelligence. (Remember the problems we had in defining intelligence in the previous chapter.)

Of course this is not the same as saying that the machine has a high IQ, because the definition of IQ makes it inapplicable to machines. But we saw also that it is unsatisfactory to equate IQ and intelligence. There seems to be no objection to saying that a machine is thinking if when it wins chess against a Grand Master it examines several alternatives to each move, and is sensitive to patterns of pieces on the board. This would be particularly true if it did not consider all the moves possible, but selected moves in relation to moves it had already made, the moves made by its opponent, games it had played in the past and the extent to which the current game resembled a past one. The only possible objection would be to say that for some reason you refuse to use the word "thinking", or "intelligence" except when the performer is alive, and a human, and that the machine is not aware that it is thinking. But why should you restrict the use in that way? Would you deny that animals think? Would you deny that self-reproducing creatures with problem solving abilities found on exoplanets (if that ever happens) can think? If you think that humans think with their minds, we have already seen that the mind is not a ghost in the body but a set of abilities.

Mental thinking is thinking that is done silently, not thinking that is done by the GIM. If you really want to say that thinking and intelligence are the properties of a non-physical entity hidden in the body, how could you prove that, and how in that case can you ever know that anyone else is thinking? It is the behaviour of the whole person that we judge as intelligent, as showing thought, as being creative. Bear in mind that we did not need to have modern scientific equipment to know when someone is thinking. Humans have known that each other are thinking for as long as humans have existed. If we are happy to allow that Daleks in the science fiction series *Dr.Who* are cunning and malign, it is because of their overall behaviour, not because we know they have a Darlekian GIM. The problem with identifying mental life with a GIM was nicely noted in a song, *"Poor Judd is Dead"*, in the musical *Oklahoma,* where the character was said to have loved the birds and the flowers, and all his fellow men, "But he never let on, so nobody ever knowed it!"

So why should we not say that computers can be intelligent and think when they reason logically? Let's return to the definition of a *Turing Machine*. A digital computer is a Turing Machine, and hence can perform any calculation that can be defined in a finite number of words, and there are plenty of pieces of behaviour that we naturally call intelligent that can be so described, such as logic, chess, and much problem solving and game playing. Furthermore, we know that humans do not only prove themselves intelligent by deep reasoning, but also by recognising patterns, triggering well-learned automatic behaviour, and so on. Furthermore, humans can show intelligent behaviour even without thinking about it consciously[173].

You may not think that a camera that can identify a human face is intelligent, but you might see intelligence in a humanoid robot that can interact with its environment, find its way about, and adapt its behaviour to the needs of people and things with which it interacts. Incidentally, the fact that we tend to think of humanoid robots as being intelligent more readily than a black box computer or a camera should tell us something about our concept of intelligence. It is not just logical ability that is important, but the way a creature shows adaptation to the demands of its environment, and even how it shows social interaction. Is a robot companion merely a camera with legs? Or could it be a friend[174]?

---

[173] D. Kahneman, *Thinking Fast and slow.* 2012.
[174] For a story in which a robot becomes a friend see, *That Uncertain Midnight,* by Edmund Cooper

We may feel that we can recognise intelligence in the reasoning abilities of computers, but do we think that such intelligence is like human intelligence? There are some computer programs that have been written deliberately to mimic in detail how humans think.[175] These programs include sub-programs that have some of the properties of human memory, the distinction between long and short term memory, limitations on the amount of information that can be held in working memory, value judgements in decision making, etc., which mimic both the abilities and the limitations on speed and capacity that we find in humans. Do we think that intelligence in other animals is like human intelligence? And what do we think the role of intelligence is in defining human nature? We need to ask this because writers such as Kurtzweil have in recent years suggested that the moment is coming when the abilities of machines will surpass the abilities of humans, and even that humans will be able to survive the death of our sun by exporting human intelligence either in the form of machines or even in the form of disembodied patterns of information across the universe[176]. Already we have automated systems that can analyse DNA, send the resulting formula to another machine, and that machine can synthesise a new DNA molecule[177]. Would we then think that we could talk of "humans" in such cases? Can we synthesise a clone at a distance? Can I synthesise myself as a clone? If not, why not? A clone is not the same person as the original any more than one identical twin is the same person as another.

Well, there is no doubt that machines can already show behaviour that exceeds the abilities of humans. So can we expect a day when all human abilities can be implemented on machines, whether computer, robot, or specially built models? How are we to decide whether there are deep limits on our ability to mimic, build, or simulate humans? We can make robots that can see in very dim lights and in

---

[175] See for example Laird, J. 2012. *The Soar Cognitive Architecture.* Cambridge, Mass. MIT Press. See also ACT* http://act-r.psy.cmu.edu

[176] R.Kurtzweil, *The Singularity is Near*, 2011. Penguin Books.

R. Penrose, *The Road to Reality,* 2004. Vintage Books. "My own view is (that) intelligence (is) the most important phenomenon in the universe." Page 362.

[177] C. Venter. 2013. *Life at the speed of light*. London: Little, Brown.

the ultra-violet or infra-red part of the spectrum. Furthermore we can in principle design a machine that can reproduce itself and hence evolve. Imagine a factory in which all the parts needed for a machine are available, and a robot assembler picks them up. It has a program that reads in part as follows, "Connect part 74 to part 128. Connect part 345 to Part 400. Now connect these two to one another." The program continues along these lines, and right at the end, it says, "Copy this program into the machine you have assembled. Connect a plug to the cable. Plug it into the mains supply. Switch it on. Cut it loose." The new machine will now in its turn make another copy. It has reproduced. The fact that the parts are available to ready-made is irrelevant. After all, our parts are made accessible to us in our diet. We can even imagine that from time to time the program joins up the parts in a new way: it shows mutations[178]. Formal treatments of self-reproducing machines were originally developed both by Turing and von Neumann[179].

We saw in Chapter 9 that we can in principle "make a human" by arranging matter in the way that is identical to the molecules that occur naturally in human sperm and ova. But that is not what we are asking here. We are asking rather whether we can build something from non-biological components that will do everything physical, intellectual and linguistic, even social, that a human can do, including being conscious. And to that the answer is "no". The reason is not because there is any special non-physical human ghost, a non-physical soul, in the human machine, because we have seen that there is no reason to think that such is the case. Rather, it is because there are good reasons to think that the limits on what Turing Machines can do prevent us from specifying the design of machines to show all human behaviour, let alone all human characteristics. Having a human body is needed to do many human things, and human nature includes embodied abilities by definition.

---

[178] As with biological mutations, very few of these would be successful or survive.
[179] J.von Neumann. 1967. *The General and Logical Theory of Automata.* In L.Jeffress. The Hixon Symposium. See also A. Hodges. 2012. *Turing, the Enigma.*

People like Kurtzweil and Penrose almost invariably identify human nature with rational intelligence, and so ask about the ability of artificial agents to solve problems, to reason logically, and to communicate linguistically with other agents. But there is more to human nature than rationality. The definition of Turing machines refers to two characteristics of what is to be simulated, namely *behaviour* and to be definable unambiguously *in a finite number of words*. But we can find aspects of human nature that fail to satisfy each of those requirements. For example, feelings and conscious experiences are not behaviour done by a human, even if they do require behaviour of nerve cells. What has feelings and experiences is a whole *person*, not a brain, or merely a collection of nerve cells. My feelings and experiences are not behaviour, not things I do, but events that happen to me. And indeed, although we can understand what it would be like to have evidence that a creature has feelings and experiences, we cannot even begin to think what steps one would have to take to make a creature *have* feelings or experiences, other than to make an exact copy of an existing creature using the existing materials. Try to write a set of instructions for an artificial network that not merely responds to the shape of a human face, but also actually has the conscious experience of seeing it. What kind of sentence could it be?

Furthermore, there are behaviours for which there do not seem to be exhaustive descriptions. Consider a concept that we studied briefly in Chapter 3, namely *bravery*. How would you unambiguously describe *bravery*, in such a way that all the behaviour described was *brave*, and no behaviour was included that was not *brave?* Since the description of *bravery* is so dependent not only on what the brave person does but also both on their feelings and on what happens in the environment, there is inherent ambiguity. Almost any behaviour can be an example of bravery, and none need necessarily be so. To run away from a threat may be very brave in certain contexts, for example if you believe that people are often "shot while attempting to escape"; but under other circumstances it is cowardice. To charge a machine-gun nest single-handed may be very brave, or may be the result of an almost pathological state of drunkenness, arousal or anger. Or consider the nature of *love*. Under some conditions it may be a sign that you love someone to cut off their foot, for example if they are trapped by the foot under a rock and flood water is about to drown them. On the other hand, to cut off their foot may be a form of torture[180]. How long must the description be to cover all possible occasions?

This kind of ambiguity is ruled out in the case of classical AI programs by restricting the problems to logical puzzles. The rules of a game are by definition precise, which is why we can imagine making a machine play chess, but cannot see how to program one to be in love. Indeed, there is much in human nature where if we try to describe the behaviour precisely we find, because of its rich dependence on context, that the description grows ever greater and more complex, rather than becoming more precise the harder we try to describe it. How can we define the difference between a confidence trickster and a person genuinely performing an act of kindness? If it is only in the motivation of the person, how can that be described in terms of behaviour? This ambiguity may be related to problems discussed by Turing and other mathematicians about the impossibility of being able to know whether a calculation will terminate, and to Godel's Theorem as applied to ordinary discourse rather than mathematics.

For many human abilities, the ability to perform cognitive tasks, the usual concern of AI and robotics, is quite inadequate to replicate human nature. One may be able to describe some of the intellectual qualities of a human that are involved in meeting, recognising, and understanding someone with whom they fall in love. But to kiss someone they must have lips; and for this to involve love it must also involve the emotions and feelings of love, and the commitment of one person to another for a considerable part of their lives, at least in our society. That in turn requires the kind of body a human has, and the chemistry of hormones and non-neural changes in the body. But above all it requires the existence not just of the lover, but of the beloved, the environment in which love is expressed, and the conventions of love as recognized in society. There is no guarantee that by transmitting the information that defines an individual's intellectual ability we can be certain that the description is one that

---

[180] W. H. Auden. *Oh tell me the truth about love*. Collected Poems. 1969.

will lead to being loved on a planet circling α-Centauri. To have a roll in the hay requires not just hay but a willing lover: but it does at least require hay!

We have seen that an individual cannot be uniquely specified just by their DNA. One would have to transmit a relevant environmental context as well. Again we are faced with the paradox that we may be able to describe the components, the atoms, but not what they do as a system. Moreover, given the inherently stochastic nature of quantum theory, it is by no means certain we could even transmit an adequate description of someone at the level of atomic or molecular structure.

It is true however that unless one asserts by definition that only living creatures can think, then we can say that AI is an example of machines that think. If the ability to calculate, to do arithmetic, and to carry out calculations using symbolic logic are examples of thinking, then AI is an example of artificial thinking, since computers can certainly perform those tasks. Someone might chose to say that AI is not really thinking, but only behaviour "as if" thinking is being used. Certainly we can point to many examples of human performance of the "as if" variety, starting with theatrical acting or the behaviour of a confidence trickster. But on reflection it seems that the main reason to say that a computer cannot think but a human can is a desire to assert that human thinking takes place in the mind, and that the mind cannot be identified with the physical brain. That is a usually a dualist's assertion, and as will by now be obvious, it is a view of human nature that is rejected in this book.

The conclusion from Turing's analysis seems to be that any function or operation that can be completely described in terms of symbolic logic can be performed by a computer. Things that can so be described, actions, even mental ones, can thus be carried out in principle by AI. To that extent mental life can be simulated or copied artificially. After all, mental arithmetic is not arithmetic done in a ghostly schoolroom in the skull: it is just ordinary overt arithmetic done quietly. To see this, do an arithmetic calculation with a pencil on paper. Now do a similar calculation without using the pencil, but speaking the words aloud as you calculate. Now do it again, but moving your mouth and tongue without speaking aloud. Finally do it with your mouth shut and your tongue quite still, "in your head". You have certainly changed from doing overt arithmetic to doing mental arithmetic, but you have not moved deeper and deeper into your mind and changed from using a Pencil as a Machine to a Ghost in the Machine! Perhaps it is a mistake about the nature of the mind, not a puzzle about the nature of thinking, that makes one doubt that AI is thinking.

There is certainly much about human mental life that cannot be reduced to the logical calculus, for example experienced feelings, the awareness of perceptions, and some inherently ambiguous activities that cannot be exactly described. These cannot in principle be programmed into machines, not because they are not material activities, but because their description is not contained within the definition of the abilities of Turing Machines. Perhaps we could in principle specify a Turing machine that would be a zombie, but not a conscious human. The situation is again like the problem of the advertisement in Chapter 3. If we restrict our language to physics we can specify the design of a neon sign, but not its contextual purpose. This does not mean that we cannot imagine coming across a nonhuman creature with a convincing mental life, or even becoming convinced, to our surprise, that a creature we have made has such a life. It merely means that we are not able to see, even in principle, how to set out to make one.

Computer Science and the Human Condition

The importance of AI, prostheses and robotics to the human condition is largely one of practical application. It is certainly true that given the almost infinite and almost faultless memories of computers, and their almost incredible speed of operation, they will soon surpass many human abilities. These will include pattern recognition and identification, the weighing of evidence in complex decisions, the recording and classifying of data and information, the management of economic and military decisions, and the supplementing and replacing of parts of damaged humans. Already so-called "expert systems" which embody human knowledge sometimes make better decisions than humans[181]. And unaided humans are not able to understand the enormously complex functions of society, economics, and government, or to foresee the consequences of many decisions. In all these areas we can expect to see AI and the simulation of nervous systems playing a role in the coming decades. But none of this tells us much about human nature, and whether the uses to which AI and robotics are put is beneficial or dangerous remains within the power of humans to decide. Bostrum's discussion of superintelligence, although not clearly expressed, raises the questions pertinently[182].

Robotics and AI are fascinating disciplines, but the deep philosophical questions about the difference between humans, other biological systems and other machines remain untouched by the mere progress of computer science. Such progress leaves us with the problem of the difference between the activity of neurons or transistor circuits on the one hand and conscious perception, thought, and feelings on the other. Why does any collection of material not just show behaviour, but also have feelings and experiences? The Fundamental Words retain their power in the face of computers. It is time to examine more closely those Fundamental Words that remain on our list.

---

[181] Shu-Hsien Liao. 2005. Expert system methodologies and applications – a decade review from 1995 to 2005. *Expert Systems and Applications, 28(1),* 93-103. P.Jackson. 1998. *Introduction to Expert Systems.* Addison Wesley.

[182] N. Bostrom. *Superintelligence: Paths, Dangers, Strategies.* Oxford University Press. 2014.

Chapter 12

Brain, Mind and Consciousness

> What is matter?
> Never Mind.
> What is mind?
> No Matter.
>
> *Punch, vol. 29, p.19. 1855*

> The mind is our beak, and the human mind is even more variable than the brain.
>
> *J. Weiner. 1995. The Beak of the Finch.*

The Mind

We come at last to the most important and the most puzzling characteristic of human nature, the relation between brain, mind, and consciousness. Today Everyday Stories almost always follow Descartes and imply that the mind is an immaterial component of human nature, a ghost in our neural machine; and consciousness is the way we experience the reality of our ghostly self, the *res cogitans* that is our *ego*. The Ghost receives information about the world by means of the sense organs, experiences emotions and feelings, and makes choices through the will, while the brain provides a vehicle for our rational thoughts and choices to express themselves in action. But it will not come as a surprise if we reject that view of human nature as misleading.

Do we really ever encounter a non-material part of ourselves? The answer must be, "no". Apart from any other consideration, how could we tell that we had done so? What would it be like? No, what happens is that we think, decide, perceive, remember. In short we exercise our various mental abilities at one time or another. There is nothing that compels me to say, "I see that dog with my ghostly mind."; "I remember in my immaterial *ego* visiting Rome."; or, "My *res cogitans* is thinking that I should invest in uranium shares." As soon as I say that it is obvious how unconvincing is such a claim. All that happens is that *I* perceive, *I* remember, *I* think. True, I am aware of my thoughts; that is, I am conscious that I think, aware of what I perceive, of what I remember. But there is nothing that compels me to conclude from conscious awareness that I have a ghost in my machine. I never meet my *ego* as such. I *myself* never meet my *self*.

I don't use my brain in the sense that I use a screwdriver to put screws in wood. I don't use my brain as a tool that is separate from me. I'm not separate from my brain as I am from a screwdriver. I *am* my brain – and of course I am the rest of me as well, and my brain is me. I am both the tool and the user. It is not because the ghostly me uses the brain as a tool, but precisely because the real me *is* my body, my whole person, with which I live out my life, including my mental life in the world, that I can have or embody my mind. Just as you could see that the real me ate a fried egg, so you can tell that the real me is thinking about something or choosing to do something. How do you do that? Well, you can always ask me and I will tell you. Often you can tell just by watching me or listening to me. And if you are running an fMRI experiment you can look directly at the brain part of me, just as you can listen to the voice part of me. What I do my brain does. What my brain does I do.

The Everyday Story of the mind emphasises what a private place the latter is. It says that only I can know what is in my mind. You and I can never know whether we really have the same experience when we look at a butterfly. My mind belongs to and houses my lonely *ego*. Or so the Everyday Story goes. But that is all highly misleading. The reason we say someone has a mind is not because we are aware of a ghost in her neural machine, but because of her ability to lead a mental life, to report mental experiences, and to perform mental actions and report on them. Neither she nor we encounter an *ego*. The reason we ascribe consciousness to someone is her ability to report and discuss conscious states and their content, and to tell us about the world, her feelings, dreams, and intentions, and because of her actions. When she reports on what she is doing we say she is conscious. Actually, that is incorrect, because we might just as well say that the reason is that she can tell us about her thoughts and memories, or to be more accurate, about what she is thinking and remembering. It does not mean that she possesses a quantity of some non-physical substance called *consciousness*. It means that she has certain abilities that are reflected in speech. As usual, let us get rid of the seduction of nouns. Why ask, "What is the mind? Where is it located? How is it connected to the body?" Let's not ask, "What is consciousness? How does consciousness reflect the world outside our head?" Rather, let's as usual frame questions such as, "To what abilities do we refer by the word "mental"? What makes us say a human is conscious?"

The Hard Problem

It is important to be clear about what is really at issue. For over 100 years we have known a great deal about what part of the brain does what. The occipital cortex at the back of the brain is needed for vision. The temporal surface of the brain contains areas that are needed for hearing and for understanding speech. Movements are initiated from a strip of tissue on the sides of the cortex. These and many other facts have been known for a very long time. As early as the 1940s it was known that stimulating certain parts of the brain electrically caused people to see flashes of light, hear sounds, or even to experience what seemed to be scenes from real life[183]. These days hardly a week passes without a report in the media to the effect that by using fMRI or some other method of recording from the intact brain we have discovered the region where thoughts occur, or religious sentiment is located, or emotions emerge. But these new reports, for all that they are fascinating, add only detail to the earlier knowledge. They are a modern version of phrenology. As a basis for the claim, often made, that they are giving us deep insights into

---

[183] W. Penfield. 1975. *The Mystery of the Mind.* Princeton. Princeton University Press.

the nature of the mind, as distinct from the nature of the brain, they are really non-starters. The real problem, what some call *The Hard Problem*, is different, and is untouched by any of this research, even the most modern. *Why does this particular collection of physical stuff not just respond physically? Why is it aware of responding?* That is the Hard Problem, the fundamental question about the mind, and knowing about what part of the brain does what gives *no* help towards an answer. The problems of neuroscience do not directly contain the answers to neurophilosophy, and one should not let oneself be misled as to where the answer to the Hard Problem lies. The Hard Problem is not to understand what part of the brain supports what aspect of mental life. That is the Easy Problem and the goal of neuroscience. The Hard Problem is to understand why the brain supports *any* kind of conscious mental life. Why are we conscious when seaweed, for example, is not? Why do hydrogen, carbon and oxygen atoms when arranged in this and only this way make something that is aware of itself, when they do not do so when they make up a lump of sugar?

Mind as a Model

Despite difficulties raised by some philosophers we have to live as if our mental life is a more or less veridical representation of the world around us: conscious thought is a reasonably accurate "mental model" of reality. We can't afford to doubt this, or it would be impossible to live. If we really thought that our perception of a solid floor in front of us might be an inaccurate representation of a deep pond, or that what sounds like the song of a lark is really a tiger's roar we would be incapable of action. Similarly if we believed that we cannot really understand what others mean when they express their thoughts verbally, we could not interact with them. We might indeed give someone a stone when they asked for bread. We must assume that our conscious mental life does really represent the real world "outside the head". But in fact we *construct* the reality that we consciously experience. Our mental life is a kind of "model dependent realism"[184]. But what kind?

The evidence that we in some sense construct our mental life is extensive. One well-known phenomenon is the "phantom limb". Following an amputation a patient may feel the missing limb to be still present. Less pathological but just as striking is the phenomenon of "projicience". When using a long screwdriver one feels not that one is turning the screwdriver but that one is turning the screw. One is located at the point of action. This occurs more dramatically in remote surgery: with a virtual reality display a surgeon feels himself to be moving not the controls on his desk, but the tip of the surgical instrument hundreds of miles away without any mechanical connection to the control. In an experiment some years ago[185] a panel with a matrix of vibrating pins was applied to the back of a blind man. The input to the pins was the amplified output from a television camera. Depending on what the camera was viewing the pattern of vibration on the blind man's back changed. Initially all that he felt were different patterns of vibration. After a while he was able to identify some of them as having individual shapes representing different objects. Finally, he reported that he no longer felt the vibrations on his back, but perceived the objects *out in the space around him.* Von Senden[186] asked congenitally blind adults to describe how they perceived space in a city through which they were walking. They reported that they were aware of a 3-dimensional sphere of space around them roughly out to a distance that they could reach with a stick, and within which sources of sounds such as vehicles moved, but that beyond that sounds simply changed in loudness without being localized in space. Perhaps the most spectacular example of how we construct perception are the experiments in the 1930s by Gestalt

---

[184] S. Hawking and L. Mladinow. 2003. *The Grand Design.* London. Bantam Books
[185] Bach Y Rita. 1969. Vision Substitution by Tactile Image Projection, *Nature, 221,* 963-964.
[186] Von Senden, M. 1960. *Space and Sight.* London:Methuen.

psychologists[187]. It is easy to design an optical system that inverts the field of vision. In one form the whole world appears upside down; in another the world is inverted left to right. When first wearing these systems the user is extremely disoriented, as one might well imagine. But after some days not merely can some people adapt their behaviour even to the extent of being able to ride a bicycle despite the left-to-right inversion, but some report that *the visual world looks normal*[188]. Thereafter, if they put the visual system on, and then took it off, they reported that the world flicked between representations to keep perception in a normal orientation. Conscious perception of reality is constructed by the perceiver.

On the other hand, perception is equally determined by the physics of the world and the neuroanatomy of the brain. We saw in earlier chapters that all nerve cells use the same kind of signal, namely an electrical pulse. Senses differ according to the origin of the signals, (eye, ear), and the destination of the impulse (different cortical areas of the brain). Although a nerve impulse in the optic nerve is indistinguishable from one in the auditory nerve those which originate in cells in the retina of the eye give rise to visual perceptions, while those from the hair cells in the ear give rise to the perception of sounds. Each sensory system gives only one kind of information. If a signal from the environment triggers the auditory receptor and the signal traverses the nerves to the auditory cortex, then the kind of sensation is auditory - we are conscious of a sound (we don't see a light). Furthermore if we stimulate the auditory cortex directly without a sensory input from the receptors we hear a sound; if we put an electronic chip on the retina and connect it up to the visual cortex we see a light. If the appropriate sensory cortex is damaged a patient will be blind or deaf.

---

[187] I. Kohler. 1964. *The Formation and Transformation of the Perceptual World.* New York. International University Press. See also http://books.google.fr/books?id=XNsDAAAAMBAJ&lpg=PA114&ots=BEC_Ata71J&dq=inverting%20spectacles&pg=PA114#v=onepage&q=inverting%20spectacles&f=false

[188] There are some problems about interpreting these reports. I give them here as originally reported.

The mental contents of which we are conscious, and their relation to external reality, arise in the interaction between the activity of the nervous system cells and the construction of our mental models[189]. We know there are some differences in the physical and chemical properties of neurons in different sensory pathways[190], but these are not differences that throw light on the "Hard Problem".

Mind and Brain

While the normal operation of the nervous system is needed if conscious perception is to occur, there are many abilities that do not need conscious perception. To look at these may give us insight into what it may be like for species lower down the evolutionary scale to be conscious, those that do not have the kind of language that humans use, and so cannot tell us about their conscious experience. Adaptive behaviour can occur without conscious perception, including what has come to be called *blindsight*[191]. We do not have to be consciously aware of what is happening in our world in order to respond to it, although if we are conscious we are always conscious of something, a visual object, a sound, a mental image, a thought. Among these are the feelings of anger, warmth, etc., and what philosophers call *qualia,* such as the sensation of *redness*. The majority of conscious states are thoughts, words, images, perceptions of meaningful objects, sounds, and identifiable feelings such as a stomach-ache or anger. There is no such thing as a contentless conscious state.

There is a great deal of human biography that requires only behavioural responsiveness. We may be quite unaware on occasions of doing quite clever things of which we are never conscious. For example,

---

[189] A few people seem to have some "cross wiring", in that they see colours in response to sounds, or perceive numbers as colored. This condition is called *synaesthesia*.

[190] Certain drugs cause deafness but not visual changes as undesirable side effects, and hence there must be chemical differences in the neurons in the two senses.

[191] N.Humphrey. 2008. *Seeing Red:a Study in Consciousness.* Harvard University Press. Cambridge.

many experiments show that the brain can produce a response that shows that it has identified information received through the sense organs even though the owner of the brain has not consciously been aware of the identity, or even the occurrence, of the sensory input. One example is an experiment in which the electrical activity of the brain was recorded while people were asleep. Signs of arousal occurred frequently when the sleepers' names were played to them even though they did not wake up, and had no awareness of the names[192]. In another experiment people heard a message through one ear while to the other ear a translation of the message was presented. If the listener was really bilingual they could not avoid hearing the translation; but if they did not know the language they did not even know that the second message was speech[193]. The meaning of the sentences was extracted before conscious perception occurred. Many other examples will be found in a recent book by Kahneman[194] on experimental studies of thinking. And some readers will have suddenly realised that they have driven a car for many seconds without any awareness of the road. We never normally hear what reaches our two ears as two messages, one at each ear, but rather a single message located in some identifiable direction in space, integrated below the level of conscious perception by neural processing. In one experiment when words were presented alternately to one or other ear in the form of phrases, the words were consciously heard on the "wrong" side if the phrases made more sense that way. The listeners' brains reorganized the sound images before the listeners were conscious of them. Blindsight experiments have shown that some humans who say they have no vision at all can nonetheless point to visual stimuli. Some blind monkeys have been able to catch thrown objects or flying insects[195]. In blindsight the external observer might want to say that the person must have "seen" in some sense, whereas the owner of the visual system would deny it: the latter is not the best judge of

---

[192] Oswald, I., Taylor, A.M. & Treisman, M. 1960. Discriminative responses to stimulation during human sleep. *Brain, 83,* 440-453.

[193] Treisman, A., 1964. Verbal cues, language and meaning in selective attention. *American Journal of Psychology, 77,* 206-219.

[194] D. Kahneman, 2011. *Thinking Fast and Slow.* Penguin Books.

[195] N.Humphrey. 2008. *Seeing Red: a Study in Consciousness.* Cambridge. Harvard University Press.

his own ability, just as an anaesthetised patient is not the best judge of his very existence! So there can be a link between the reception of information by sense organs and adaptive overt behaviour which shows that the brain can respond adaptively without any conscious awareness by the person who responds.

Now all the above facts relate to a story about why a sensation or perception is linked to a particular neural pathway, to a particular part of the brain. It does not say why there is such a thing as sensation or perception. We are back to the Hard Question. Moreover, in most writings about the nature of consciousness there is a tendency to concentrate on perception. But there are other aspects of being conscious that do not raise directly the question of how we are aware of the external world. I can think consciously about a problem or puzzle without using perception. I can reason, or translate a remembered poem in my thoughts. I can reflect on my memories of my travels last summer. I can dream.

What account can we give of the relation between overt behaviour, conscious mental experience, and neural events? Neuroscience does not actually give us privileged access to the mind. If all we had were records of the electrical activity of parts of the brain, and if no patient ever told us what experiences she was having at the time the electrical signals were recorded, we could never know what the signals represented. We still have a great mystery. Stones, coal, rhubarb and carpets never show any signs that they are conscious. What is it about many living things, and humans in particular, that makes them able to show evidence that they are conscious? We saw earlier that the atoms that make up living things are no different from the atoms in non-living things. There is nothing special about the chemical composition of nerve cells: potassium is potassium, carbon is carbon, hydrogen is hydrogen. So why is it that when chemicals are put together in the way they are in nervous systems, we find that people are conscious and aware of what is happenig as well as showing overt behaviour?

Let's start by asking what kind of answer we should be looking for. Does it make sense, for example, to ask for a causal scientific explanation of the form, "consciousness is caused by the properties

A, B, and C of nerve cells, which in turn are due to the properties D, E, and F of carbon atoms, which are due to the properties G, H, and I of subatomic particles such as quarks."? The answer must be, "no", because we have already seen that there are no such special properties that would let us tell such a story. Would it be sufficient to say that what is needed is a complex network connected to memory and fed by sensors that pass information from the external world, but the composition of the network does not matter? After all, we saw that we can imagine replacing neurons by artificial electronic circuits and still expect normal function. But why should the connectivity of a network result in awareness? Or dare we say there is no explanation, or perhaps the situation is like the fable of the advertisement in the Chapter 3? That would seem almost a gesture of despair.

There are two ways of talking about mental events in a conscious creature. First, you can ask what is happening by putting a verbal question to a human, and you receive a verbal report about the state of the person as experienced by that person. On the other hand you can put a question to the person's nervous system by recording the electrical activity of the brain, or by watching a fMRI. In that case what you receive is an answer about the nervous system, not about the person. Reportable conscious mental life is a property of a human being, not of a brain, even if the nervous system is the vehicle of consciousness. Brains don't tell you about their experiences: people do. Brains can only tell you about their electrical activity. So what kind of explanation do we want? If reportable mental life is not a physiological process, just as the advertisement was not an electrical phenomenon, then electrical events in the nervous system can't explain it, although they are needed for it. But of course, as with the advertisement, to say that mental life cannot be described in electrical terms does not mean it is immaterial in the sense of a GIM. If consciousness is a way of referring to the fact that there are some special ways we use language to describe our existence, then it is by looking at how we use language that we will get a better understanding of consciousness, not by recording electrical signals.

Exorcising the Ghost

Let's look back at what Descartes said, because after all his Story about dualism is the dominant Everyday Story of our time. For Descartes the most important thing about consciousness was that it was central to his account of how I can be certain I exist: "I think - therefore I am." Descartes was correct to say that if I am aware that I am thinking then I must exist. But that does not tell us anything about the nature of "I". He went on to say that his nature: his *ego*, must be a *res cogitans*, a thing that thinks. This does not prove that thinking can occur in the absence of a body, merely that he could imagine not having a body and yet be able to think. Whether that is coherent is another question. In fact there is some doubt about exactly what he meant, since Geach and Anscombe[196] in their translation of Descartes maintain that at the time the latter wrote, *res cogitans* meant "a conscious thing," not a "thinking thing." And perhaps Descartes was wrong to think of the *ego* as being a "thing" at all.

Even if Descartes's arguments were correct, all they show is that there could be occasions on which one would be certain that one was thinking even though one was at that time uncertain as to whether one had a body. It does not show that one is able to exist like that always (as a ghost) or that most people are like that most of the time. (Aquinas for example maintained that a soul is not a person.) Nor does it show that as Descartes maintained physical bodies (e.g. other animals or machines) can't think: merely that in principle non-physical "bodies" *can* think. But even then there is the problem that he has imagined away other people so that only "he" exists: but where did the language in which he thinks come from?

Imagine my existence if I were a truly lonely *ego*. I have an experience, and I call it "X". I think to myself, "I just saw an X". I have another experience and say to myself, (or rather think to myself, because by definition I don't have a body and so I can't say anything), "I saw that X again". Even if I consider the case not of an abstract "X" but, for example, a "cat", how could I ever be certain that I was correct? How

---

[196] P.T.Geach and E.Anscombe. *Descartes: Selected Writings*.

could I check that the second experience was actually the same as the first? And how did I learn to use the word *cat* for this experience, and how to use the words, *that* and *again*? How did I learn the grammar and language with which I think? It makes no sense to talk of a lonely ego using language, because a truly lonely ego has no experience by which the coherence of language can be developed in the way it occurs for humans in their social interaction with one another. It is of the essence of language to be social, not private. Furthermore, how can I see anything if I have no senses?

There is another, I think even more obvious problem about Descartes's dualism. It was meant to show that the state of consciousness was not tied to the states of the physical body. But anyone who has had deep anaesthesia for a surgical operation knows that while we may not have direct proof that thinking can occur without a body, we certainly have evidence that the living body can occur without consciousness! What, after all is the typical experience of deep anaesthesia? You are aware of your surroundings, you feel a little fuzzy or warm, and then... You are aware of your surroundings again but they are different and the time is different and your state is different as you wake from anaesthesia. Far from being able to access your ego's state of consciousness independent of your body's state, you are not aware of existing at all when you are anaesthetised. You are not aware that you exist, and you are not aware that you are alive. The surgical team knows better than you do whether you exist. But you certainly would not want to say that you stopped exisiting while you were unconscious! All this is hardly what you would expect from a Cartesian ego. If your conscious mind, your ego, is meant to be able to exist when your body dies, why does it become completely inaccessible to you when your body is still alive, just because a bit of the brain has been chemically made to change its state? Princess Elizabeth, one of Descartes's pupils raised this question with him, and he seems to have given no satisfactory reply. And why does dementia slowly destroy reason? As Aristotle pointed out, "... in old age the activity of mind or intellectual apprehension declines only through the decay of some other inward part." A Cartesian ego should not experience dementia.

Other Minds

While we justify the claim that another human is conscious by referring to verbal behaviour, we can also infer it from non-verbal behaviour, because we can relate the nonverbal evidence of consciousness to verbal evidence. There has recently been a flurry of interest in a report that one can identify what a sleeper is dreaming about from electrical patterns in the brain. But the media reports failed to notice that this was possible only because the researcher woke the sleeper and asked what they had been dreaming when a pattern was recorded. Then a recurrence of the pattern could be used to identify a recurrence of the dream. That is hardly worthy of note. It is the same thing that happens when you identify someone in a recent photograph after learning from an earlier photograph what he looks like. If all you have is a record of the electrical activity, and there has never been an occasion when a sleeper reported the content of their dream, you will never be able to know what is represented by the electrical activity.

We can even make a case that I can be as certain about your mental state as I can be about mine, without listening to your report. First, remember that there are many occasions when what I am mistaken as to what I think I perceive in the world around me. The owl I saw in the barn turned out to be a sack of grain. The bull I saw in the foggy field turned out to be a rock. Awareness in one's senses can be fallible. Now consider two occasions. In the first I am shooting grouse with a friend. I assert with complete certainty that he has seen a grouse, that the content of his mind is a bird in a nearby bush. How can I be certain? Because I see him suddenly turn, raise his gun, point it at the shrub and pull the trigger; and a moment later I see a bird fall out of the shrub. What possible evidence could make me more certain about the content of anyone's mind, even my own? Or again, I am out on a patrol in the army, and I see my companion suddenly spin to the left and raise his weapon. He calls out, "On the left! Shoot!". I look where his weapon is pointed and see the shape of a man pointing a gun at me. Am I less certain about what he sees as he fires than I am about what I see, especially when I can remember many occasions when I made mistakes in identifying what I was perceiving?

Even so, there is a sense in which my lonely ego cannot be wrong about the content of its experience, even though I may make a mistake in judging what the state of the world is that I perceive. If I am sure that what I am currently conscious of is a spaniel, then that is indeed what I am currently conscious of, even if there is no spaniel in my visual field and I am hallucinating. If sensation is always correct that is not because it occurs in a secret, nonphysical place. If I showed you a spectrum with what to me is the red area illuminated and you systematically said it was blue, it would not necessarily mean that you have a different sensation: it could be that you have not learned how to use colour language properly in our culture.

Now just as nonverbal behaviour can be justified as evidence of conscious experience or feelings, so can physiological data such as fMRI or direct electrical recordings. We can come to say not unreasonably on the basis of physiological recording, that someone is perceiving a light, having a dream, even thinking religious thoughts or making a decision: but we can only do this because, at some time or other, we have had a verbal report from someone reporting that, "I am seeing red", "I am feeling a pain", "I am saying my prayers" when their electrical brain activity was being recorded.

By analogy we can find nonverbal evidence for consciousness in animals other than humans. For example, when the occurrence of visual perception is asserted by a human it is typically accompanied by orientation movements, eye movements, prolonged fixation on visual objects of interest, and so on. Tactile consciousness is accompanied by stroking, scratching, rubbing, moving the fingers over the surface of interest. But remember that always, "evidence for the presence of consciousness" means neither more nor nor less than "the kind of behaviour that means that the creature is thinking, having experiences, perceptions, feelings, etc.". It does not mean that the creature possesses a mysterious non-physical substance. This is what I think Rowlatt means when she says, "phenomenal consciousness (is) what it feels like to experience, or be aware of, a variety of external and internal events[197]."

---

[197] Personal communication.

Science, Cells And Souls

If I can learn to tell from other humans' non-verbal activity what they are conscious of, I can do the same for dogs, or monkeys, or. . . But of course common sense tells us that. Anyone who owns a dog will be well aware that the dog is conscious of all sorts of things, many of which are not available directly to me as a conscious human. As the terrier Quoodle says[198]

> They haven't got no noses,
> The fallen sons of Eve;
> Even the smell of roses
> Is not what they supposes;
> But more than mind discloses
> And more than men believe.
> . . . .
>
> The brilliant smell of water,
> The brave smell of a stone,
> The smell of dew and thunder,
> The old bones buried under,
> Are things in which they blunder
> And err, if left alone.

When we watch another species of animal we see behaviour that informs us of its state of consciousness just as well as a verbal report informs us about humans. We see the eyes point first in one direction then another. We see the ears twitch or prick up, the nostrils dilate. We see a paw withdrawn instantly from contact with a hot or pointed object. We don't see any evidence of linguistically supported thinking, but we certainly see evidence of dreaming; and just as we can sometimes think by manipulating images in the mind, so, we must assume, can other animals.

How far down the evolutionary sequence can we detect conscious behaviour? Several philosophers have followed Nagel's suggestion that to say something is conscious is the same as saying that there is something that it is like to be that creature. He famously entitled one

---

[198] G. K. Chesterton, *The Flying Inn*. London. Methuen. 1914

of his papers, "What is it like to be a bat?"[199], drawing attention to how different the conscious experience of an animal inhabiting an auditory world must be from ours, a conclusion with which Quoodle would certainly agree! And as Wittgenstein said, if a lion could speak we would not be able to understand him. When we try to imagine what consciousness is like in another species, what it is like to be another animal, it is hard to avoid thinking of a world of human vision, so visually dominated are our lives. But think for a moment of an earthworm you dig up when gardening. Its body surface is supremely sensitive to touch, it writhes when damaged, it strives to find its way back under the soil. Perhaps even an earthworm is conscious, in its impoverished wormy way, of its world of touch. (It does have an, admittedly small, nervous system.) Perhaps even simpler animals such as coelenterates like jelly-fish and Hydra have some rudimentary experience in a way we cannot even begin to imagine. After all, unicellular organisms move away from noxious chemicals, seek out appropriate concentrations of oxygen, and so on. But of course Lloyd Morgan's Canon warns us to interpret movements of unicellular organisms only in terms of the physical properties of cell membranes, unless we have very strong reasons to the contrary. And we should be very careful not to impute more than we can justify to the mental life of any creature.

It may be that where we draw the line as to what creatures are conscious is a matter of where we feel we can understand and empathise with their modes of existence, rather than an objective matter independent of the observer. But if a worm is minimally conscious we do not want to say that it has a wormy ghost in its machine, a vermicular GIM. If we ever met creatures on another planet, made of chemicals other than those found in living creatures on earth, and with completely different "nervous systems" from ours, what would make us decide that they were conscious? It could not be because we could record the electrical activity of their brains with an fMRI system, because they might not have either the physiology or anatomy to let us do so. But we could still become convinced

---

[199] Nagel, Thomas. What is it like to be a bat? », *The Philosophical Review*, Vol. 83, No. 4 (Oct., 1974), pp. 435-450.

by their behaviour, even if they are made of metal and clay, and with petroleum blood systems, that they are aware of what goes on around them. We would not know that they are conscious because we have observed their non-material mind or consciousness, because no such things exist. By their behaviour we would know them to be conscious, even if we could not empathise with what it is like to be them. If that were not true, people could never have known that other people are conscious until electrophysiological recording had been invented, which is obviously absurd. Because I know how a camera is made, I know it cannot be conscious: but in the case of exoplanet aliens, although I do not know how they are made, I might come to believe them to be conscious. Certainly if they had a language comparable to human language and we learned it, we could become certain that they were conscious.

The Mental Life of Plants: a fable.

In my garden there are two unusual plants[200]. The first is *Mimosa pudica*, the "Sensitive Plant". If a heavy shower of rain hits it, or if you tap its leaves smartly, it folds up its leaves and droops like a collapsed umbrella. After some time, when the rain has stopped, or it has not been struck again, it comes upright and reopens its leaves. The second plant is the *Venus's Fly-trap*. When a fly or certain insects touch the leaves, the latter in an instant close up imprisoning the insect, which is then dissolved by juices from the plant, so that the plant is nourished. If you drop a small pebble onto the leaves, they do not respond, so they can discriminate some kinds of living from nonliving material.

Do you think these plants, which of course have no animal cells, and no nervous systems, are conscious? If so, why? If not, why not? In fiction there are at least two well-known examples of conscious plants, triffids[201] and ents[202]. In both cases although they have no nerve cells the narrative makes it plausible that they "feel"; and the

---

[200] Plants such as these really exist - but not in my garden at present.
[201] J.Wyndham. 1951 *The Day of the Triffids*. Penguin Books.
[202] J.R.R.Tolkein, 1954. *The Lord of the Rings*. Volume 2.

triffids are quite plausibly described as having sense organs and communicating. Tolkein's ents are plausible precisely because they talk. On our planet we do not ascribe consciousness to things without a nervous system: but on other planets might it be different?

Me, Myself, and I: Aspects of Consciousness

How and what do I know about my own consciousness? Obviously I don't depend on information about the electrical function of the brain, otherwise no one could have known about their own mental state until about the middle of the 20th century! It helps to talk about this in the way that Kenny[203] does, namely to distinguish between vehicle, symptoms, and criteria. The vehicle of conscious experience seems to be the physical structure of the brain. It is the material structure that supports the events that are necessary for experience. The symptoms of experience are the electrical and chemical events that, when we record them, make us external observers expect that the person being observed will report the occurrence of experiences. The overt behaviour of other people and animals are also symptoms of mental life. The criteria of experiences are the abilities of the person being observed to report what he or she has experienced. I know about my experiences because I have direct access to the criteria. I do not deduce my conscious state from any symptoms.

I can be as certain that a hunter or gunner has seen a target as that I have. If I judge that I have seen a grouse I am no more certain than of the fact than that I have seen the gunner pull the trigger. Consider the model-determined-reality (MDR) analogy for seeing a target. MDR complementary models yield complementary descriptions which are based first on awareness by oneself, second on observation of the behaviour of another, and third, recordings from the brain. But as in the fable of the Electrician and the Advertisement, these are all different ways of describing the same thing.

I clearly don't become aware of the contents of my mental life by having access to the symptoms of my consciousness. So what are the

---

[203] A. Kenny. 1989. *The metaphysics of mind.* Oxford. Oxford University Press.

criteria I have about my consciousness - surely just that I do in fact see (touch, hear) things? In the experiment in which people responded to their names in sleep the criterion for saying that the person heard their name was basically that they told the experimenter, but they nominated nonverbal behaviour as a proxy, after which that proxy could be used. Compare this with the ultrafast involuntary response of a gunfighter, who may learn that he has drawn his gun and fired from seeing himself do so, not from consciously deciding to do so[204]. Neuroelectrical events are symptoms of consciousness. Someone's experiences, about which the person can tell you when asked, are criteria. Verbal reports are criteria of consciousness, providing we assume honesty in the speaker.

One thing that perhaps makes it harder to understand the nature of being conscious is the dominance of vision in our lives. Vision emphasises the distance of the external world from the hypothetical lonely ego. It seems obvious that the world is "out there" and my perception of it is "in my head" as if it is a picture, separated from the thing perceived. We tend to think that the content of our awareness is not the "real thing" as it exists in the outer world, but a picture in the mind. That in turn makes us believe that the picture must exist in some inner "mental" projection booth. It may help to dispel that idea if we consider other senses. If we listen to music from a pair of stereo speakers we do not hear the music in our heads, but "out there": if we alter the balance control on the stereo system we hear the source of the sound move across the room. If we listen through headphones, however, we hear the sound move between the headphones, and then indeed we hear the sound in our heads but not in some ghostly non-physical sound studio. Our experience of sound is related to where the source is in space, not in a GIM. Think about touch. If we are touched, or stroked, or struck, it is obvious that what we experience

---

[204] I once took part in an experiment on reaction time in which we undertook intense practice for hundreds of trials to increase our speed of response. In the end, one found out which response one had made by observing what one had done, not by deciding to do it, so rapid and automatic had the behaviour become. See Davis, R. Moray, N. Treisman, A. M. (1961). *Imitative responses and the rate of gain of information.* Quarterly Journal of Experimental Psychology, 13, 78-89.

is the actual world, not a displaced image, and we saw that Bach-Y-Rita's patient constructed an external representation of the vibrating pins on his back. There may seem to be a gap between the viewer and the viewed. There is no such gap between the toucher and the touched or the listener and the heard. And with that realisation the temptation to think of internal ghostly images disappears. When we are conscious, we are conscious of reality, and it is with reality, or a model of reality that our thoughts, memories, and other mental abilities are concerned.

The Origin of Consciousness

Well, where have we come to? Consciousness is not a thing, a substance, a part of me, not a pattern of electrical activity in some part of the brain, but rather "being conscious" is a phrase to describe the fact that I can tell you about the world around me or about my thoughts and memories when my brain cells are stimulated by energy and chemicals. I can make use of language to reason, and I can be aware of choices I make and report about them. Because of language you can know, often as well as I, what I am thinking or perceiving, and even what is the content of my imagination, my mental images and my dreams. And from knowing such things I can learn how to tell what someone else is consciously doing by interpreting their non-verbal behaviour. For conscious perception everything must work - retina, optic nerve, cortex and reticular system - so there does not seem to be sense in asking for the seat of consciousness, but only, under what conditions does the nervous system provide symptoms and criteria. Ask the same about imagination and dreams.

The question remains, why does this particular congeries of physical stuff not just produce behaviour, not just provide symptoms, but actually allow the realisation of criteria? Perhaps it is true that there are parts of the brain which if removed, or if their action is suppressed by chemicals, suppress conscious mental events; and in such a case we might say this part is needed for conscious mental life, although even then it is not really the "seat of consciousness". Consciousness is not something that can be located. It is something experienced. With a mechanical vision system there is no difference between symptoms (the electrical currents) and criteria. From animals we obtain only symptoms, including behaviour. Because humans have language there is the possibility of a break between symptoms and criteria, we can report on criteria, and it is that fact that makes us think there must be an immaterial mystery.

This may seem a surprising conclusion, and indeed almost as if we are abandoning a rational approach to the nature of being conscious. If so, does this mean that being conscious is unlike any other topic in science? Well, think back to the fable about the Electrician and the Advertisement, indeed to all we have said about complementary languages. To say that there is more than one way to describe a single event, in this case the conscious experience of the world by a person, is just to repeat what we said then. Remember, having more than one account of an event does not mean, necessarily, that one account explains another or precludes another. So to say that an event can be described in several ways does not mean that we are abandoning rational discourse. No indeed, for the deepest and most fundamental statements of theoretical physics take such an approach to their subject matter. Consider what scientists have said about the quantum physics, in particular quantum electrodynamics.

> ... nevertheless, the answer to this dilemma that emerged in the early part of the 20th century, primarily from Einstein's speculation, was that light is indeed both a wave and a particle, depending on the way that it is studied in experimentation. This idea became known as "wave particle dualism".
>
> Mendel Sachs[205]

And when discussing the disturbing feeling one has about quantum physics, that surely there must be something deeper to explain why the mathematics of quantum theory describes the world so accurately, the physicist Richard Feynman[206] said,

---

[205] M. Sachs. *Einstein versus Bohr.* 1988 Open Court. Lascelles Illinois. Page 78.
[206] R. Feynman, 1985. *QED: the Strange Theory of Light and Matter.* Princeton. Princeton Science Library.

> It's no good asking what there is deeper down. There is nothing deeper. All we can say is that this is how the universe is at its deepest level of description.

If that is permissible as an approach to the most profound theories of physics or to the existence of gravity, perhaps we should take the same approach to being conscious. Being conscious is a property of living organisms, or at least some of them, such that if you ask questions using a method that is appropriate to studying the electrical activity of the brain, then what you get is an answer in terms of electrical activity. If you ask questions in a way that is appropriate to obtaining reports about the experience of the whole organism, then what you get is that sort of report. To study the underlying vehicle, as the philosopher Kenny would call it, you use electrical methods. To study the organism you ask the whole organism what is happening. To do the latter requires language in the first instance, although with language you can establish an interpretation of other kinds of behaviour, including neural behaviour, that generalises to non-language-using creatures. So a conscious report and a report of the electrical activity of the brain are indeed about the same thing, but are two aspects of a single reality, in a similar way to wave and particle being two aspects of the same thing, radiation, in quantum physics. It seems we have to settle for what we may call, by analogy with physics, the dual aspect theory of consciousness. Which picture of reality is taken as primary will change from occasion to occasion, as described in Model-Dependent-Reality by Hawking and Mlodinow[207]

Of course none of this implies that consciousness requires a person to have a non-physical GIM. It is clear on the contrary that being conscious of the world around me, that is to consciously see, hear, or touch some aspect of the world needs a body. So does thinking or remembering. Consciousness is not just a state of the central nervous system that occurs when some central part of the brain "lights up" in an fMRI experiment. On the contrary, to see a dog I need a dog, light, an eye with its retina, receptors and nerves, the optic nerve

---

[207] S. Hawking and L. Mlodinow. *The Grand Design*. 2010. Bantam Press. London

taking signals from the eye to the brain, and all the deep structures of the brain. We should not ask for the seat of consciousness, but only under what conditions is a person conscious. Nothing suggests that there is a non-material element in this mechanism. Indeed the opposite is the case. If the eye is missing, or if the central nervous system is anaesthetised, there is no visual perception. Only if all the physical system is working is there conscious perception. That is why I have had to adopt a two-language approach to understanding how we should speak about consciousness.

We need a dualistic theory of consciousness to describe human nature, but it is not dualistic in the sense of Descartes. His was an *ontological* dualism. He really thought that there were two components of a human, made of different kinds of material, one physical and one not physical. We have to be careful when talking about neuroscience, because sometimes people talk as if there is a difference between my brain and me. If my motor cortex initiates action before I am aware of deciding to act then, it has been suggested, it cannot be I who acts. But as we shall see in the next chapter, that is a mistake, for my motor cortex is part of *me*.

What we need is a *logical* dualism, two ways of talking, not two components of being[208]. Chalmers[209] suggests that just as the quantum laws of physics reflect Fundamental Laws of the world, so for living creatures there may be a Fundamental Law that says that certain configurations of matter are vehicles of conscious experience. If that is so, it would be no more arbitrary a fact about the universe than the facts of quantum theory or the existence of gravity.

We should not let ourselves be scared by an apparent lack of causal explanations. Such a situation is not uncommon in the history of science. We have just seen how quantum physics strands a questioner at a point where there simply is no more profound explanation. All we can say is, "That is how the World is." In Newton's lifetime a major objection to his proposed laws of gravitation was of this kind.

---

[208] Rowlatt calls this second way of dealing with the topic a *metaphysical* dualism, which is a good way of putting it. (Personal communication.)
[209] D.Chalmers. 1998. *The Conscious Mind.* Oxford. OUP

People objected to the fact that he invoked the notion of *force* that acted at a distance across empty space with nothing to let one body touch another. His laws simply said, "This is how nature is." In our case being conscious when matter has a certain configuration is "how nature is." In a mechanical vision system there is no difference between symptoms (the electrical currents) and criteria. In animals there are symptoms, including behaviour, and criteria for the animal in the form of successful adaptive behaviour. Because humans have language there is the possibility of a distinction between symptoms and criteria. Looking at physiological responses on the one hand or asking for a verbal report on the other is equivalent to setting up wave experiments on the one hand or particle experiments on the other in the quantum physics dual-slit experiment. "That is how nature is."

The existence of conscious awareness in living creatures is, then, a mystery, although not one involving a ghost. But many people think that there is a deeper mystery in the existence of *self-awareness*, or *self-consciousness*. What is the nature of the *self*? To think like that is a mistake: one deep mystery is enough! The belief that there is something special about self-awareness stems from the mistaken belief that the noun *self* stands for a special non-material part of the person, roughly equivalent to the *ego* that we discussed earlier. As we have seen, we can tell that many animals are consciously aware of their surroundings. Anyone who has taken a dog to a vet for an anaesthetic, has seen the anaesthetic administered, and later watched the animal reawaken will know that one can tell when animals are aware of the world around them. But humans are different. They are aware of their own awareness. They are conscious that they are conscious. It is indeed that observation that is central to Descartes's argument. I am not just aware of the world around me, of things that stimulate me, of what I do, but I am aware that I myself do them. What should we say not just about consciousness, of which we have been talking until now, but about *self-consciousness, self-awareness*?

Here again, perhaps here above all, we must be wary of the seductive power of nouns. We were warned in a quotation from Kenny in Chapter 3 not to be misled by the grammatical difference between *myself* and *my self*. I share conscious awareness of the world with many creatures, with spaniels, flamingos, and sharks. But only a human has a language that let's her consciously report about herself. So what is it that a human is aware of in such a case? It is *not* a self, an *ego*. That, as we have seen repeatedly, is a Cartesian illusion. A human, in addition to the fact that she is aware of the world is aware that she is aware. That is the central meaning of self awareness. And what does it mean to be aware of being aware?

Let's repeat, to make sure we are clear, that it does not mean that one encounters a GIM, an immaterial ego, an immaterial self. The truth is much simpler, and again is a matter of our relation to language. No other kind of animal has a language with the grammar and syntax of human language, and it is impossible to overemphasise the difference that makes to how humans confront the world. If I claim that I am self-aware I am not claiming anything other than that I am consciously aware of the world, consciously aware of what I feel, what I think, and what I do, *and that I can talk about these things while correctly using the first person singular of verbs.* When I report the conscious awareness of other animals or other people I report what can be concluded from the symptoms of their behaviour, and my human language allows me to make my report using the second or third person of a verb: "You must be feeling hot."; "The dog has just bitten the cat." By observing the behaviour of others I have access to the symptoms of mental life, whether conscious or not, in others; and by using the 2nd or 3rd person form of verbs, I can correctly report their mental life. When I want to report the criteria of my own mental life I use the 1st person singular of the verb: "I have just had an idea. I can take the dog for a walk in the park I see outside."

To claim that I am self-aware is the same as to claim that I can correctly use the 1st person singular of all the verbs in my language. And don't think that is a trivial claim. Human nature depends on it. If I am speaking English I say, "I feel cold." Cicero might have said in Latin, "Frigidus sum." In Greek Aristotle might have said, "Cryos eimi", "κρυος ειμι". The fact that a person can *correctly* use the verbs of his language in this way is proof of self-awareness. It is indeed what self-awareness means. This ability is a gift of evolution that humans received at the time that human language was added to the repertoire of human abilities. With that gift came the ability to be self-aware. Self-awareness is not the addition of a ghost to our neural machine. It is the addition of the ability to make first-person reports about perception, feelings, thoughts and intentions. As with so many things, the most important difference that marks of humans from other animals is the nature of human language. I don't want to insist that the only way a person can think is by using symbolic language: perhaps one could also think by manipulating images in the mind's eye, so to speak. I think we could call image manipulation in our imagination a way of thinking, and this may be available to other animals. There is also evidence that some human problem solving is not the result of thinking but of checking present perceptions against a huge collection of memories for past events until a match is found.[210] But the typically human, characteristically human, way of thinking, the way of thinking that marks us off from any other species we know, is the use of symbolic language, and that is what we mean by using our mind.

If you feel the need to make your *self* more secure, or a psychotherapist says that your *self* needs to be more fulfilled, what they are saying is that the way that you talk shows that you are uncertain how you feel about life, not that your GIM is ill. Hopefully therapy will improve the way you look at life. But remember that your self-image is not an image of your *self*. Your self-image is just the collection of your habits, your opinions, your ideas, and your feelings as known to you. It is not a picture of an immaterial ghost inside your skull. It is the total account of you as a whole person. It cannot be found in a fMRI image because the latter is not a person.

In fact the notion of a *self* seems largely superfluous. It is interesting to compare the syntax of French and English with regard to *self*. In English we start from *myself* and develop the notion of *my self*, or *the self*. We then think of the *self* as being some kind of psychological component, a part of my psychological makeup, and something that contains or displays my identity. But in fact we never encounter our *self*. In French, we start from *moi-même*, and develop the notion of

---

[210] R.Beishon. *An analysis and simulation of an operator's behaviour in controlling continuous baking ovens.* In E. Edwards and F. Lees (eds) *The Human Operator in Process* Control. London. Taylor and Francis.

le moi, not of *le même*. That is, the French emphasises the person, *me*, not the (phantom) *self*. There seems to be no scientifically useful sense of *self* that cannot be handled just by using the word "*I*" with an appropriate verb. The only quasi-scientific usage of *self* in English is in areas such as psychiatry and psychotherapy, and no modern research implies the need for such a concept as the noun implies. It is noteworthy that when people write about neuroscience research such as that using fMRI or other ways of recording electrical activity in the brain, they never suggest that they are examining the physical basis of the *self*, (or for that matter of the *soul*), but rather the *mind*.

While many sophisticated people are happy to ignore the *self*, and to say that they do not believe in the existence of the soul, not even unreconstructed Behaviorists often deny that they have a mind. The reason for this is clear. It is obvious to each of us that we have a mental life in some sense of the words. Our analysis of the meaning of mind shows that we require the concept to describe the life we know we live. We know that we perceive things, recognise people, think about things, reason, learn and remember skills and events, even decide to perform a jump, - in short, that we have what we are happy to call our mental life. There is no clash here with scientific research. We have brains and nervous systems and bodies that develop under the direction of genetic mechanisms interacting with the environment, and it is with those bodies that we live out our lives, including our mental lives. Once we exorcise the Ghost from the Machine we can feel at ease when science gives us evidence of how the (neural) machine provides a vehicle for our mental lives. We do not *have a mind*; we *live a mental life*. And science can explain how we do that just as it can explain how our muscles, digestion, and respiration work. The only thing that can cause difficulties in combining science and philosophy in our quest to understand mental life is if we allow ourselves to give in to the call of the Cartesian sirens sitting on the rocks of metaphysics and singing their dualist songs.

Chapter 13

Free Will and Responsibility

> There was an old man who said, "Damn!
> It is born upon me that I am
> An engine that moves
> In predestinate grooves,
> I'm not even a bus, I'm a tram."
>
> *'Maurice Evan Hare*

It's time to go back again to Saltarella. In Chapter 2 we saw several accounts of how and why she jumped. We looked at different causes, from subatomic to historical events. And we saw that there are many possible accounts, each of which may in some sense be true, and which complement each other. But we omitted one very important way of talking about what made Saltarella jump, the notion of "free will". She *chose* to jump - or was she pushed?

People like to think of themselves as free to make decisions and act without any cause other than their will. They want to feel that there is nothing pushing them around, forcing them to do things. There is something, a part of them, with which they make decisions and act. That something is called their *will*. It is almost like a tool that is used to carry out their wishes. But is it really true that the will is free to make autonomous choices? Many people have doubted it.

Traditionally the will was one of the powers of the soul, and so a live human being could be said to have a will, indeed a will of their own. But what exactly did that mean? We have often seen in this book that there is a tendency to think of the human soul or mind as made up of different parts of nonphysical machinery. So the will seems to be a component, or a device, a bit of psychic or nonphysical machinery which I use to make my decisions. But we've already seen that there are serious problems if we talk like that. There really is no reason to think that I have nonphysical components such as a GIM. And even if there were, it's very difficult to see how I could use a nonphysical part of me to affect a physical part of me and so cause action. Do I first decide to do something and then sit back and watch my body carry out my wish? Do I first think out everything I am going to say and then activate the muscles of my mouth voluntarily to express those thoughts? Do I always do everything twice, first willing it and then carrying out my intent? Obviously not. So what exactly is the point of talking about having a will? It is interesting that Aristotle seems not to have had a word that is equivalent to the English *will*, although he was certainly interested in morals and ethics. Aquinas, using Latin, could use *voluntas*.

Did Saltarella jump because she wanted to, or because she had to? The question of whether we are really free to chose for ourselves, or whether freedom is an illusion and we are forced to do whatever we do is one that has worried people for as long as philosophy has been practised. Even today there is a flood of books and papers about the nature of freedom and whether it is an illusion[211]. In this chapter we certainly can't review all the ideas that have been proposed. I will discuss just a few, but those will, I hope be sufficient to indicate at least one way to justify our ability to chose, and to show that we can claim a certain kind of freedom for moral actions; for it is their moral freedom, their responsibility, rather than whether they are free to put one foot in front of the other that really concerns people.

We worry whether things that we do are really *our* actions, or whether we can't help doing them. That is the problem of free will. There are many things, small everyday actions, that we do which are quite trivial, and we would not worry one way or the other as to whether we are really responsible for them. We pick up a piece of paper and throw it into a wastebasket; we start with one foot or the other when walking down a flight of steps. But while we usually feel as if we choose what to do from one moment to the next, we have to admit that there are many potential causes for our actions. There are the physical structure and function of nerve cells. There are physical

---

[211] G.Watson. 2003. *Free Will*. Oxford University Press.

events in the world around us. There are our habits, the result of learning over many years, which predispose and perhaps force us to do things because we have been "conditioned" to do so. There are even inherited traits and characteristics in our DNA. There is social pressure from our peers and friends.

It is pretty obvious that many actions in the life of a human are *not* in any sense free. For example, the movements of the pupil of the eye in response to changes in illumination, as it opens and closes to keep the illumination on the retina approximately constant is not under voluntary control. The secretion of hormones into the blood, the actions of the kidneys in filtering the blood, and events at the level of individual muscle cells, or the transmission of electrical signals in nerve cells are not normally under our direct voluntary control. But that doesn't make us feel that we are not free.

It is interesting that under special conditions some of these normally automatic actions *can* come under our conscious, voluntary control. For example, in the 1960s experiments showed that people could be taught to twitch a single muscle cell voluntarily. Usually we don't do this – we just make a movement. In fact, rather than using a muscle cell to make a movement, we use a movement to make a muscle cell contract. But if an electrode was poked through the skin into a muscle and embedded in a cell, and if the electrical activity of that cell was amplified and displayed on a screen to the person taking part in the experiment, then he or she could learn to make just one cell fire. Initially when they were asked to make a small muscle action there was a huge burst of activity from many cells. But by watching the display, and gradually making smaller and smaller movements so that the level of activity on the display became smaller and smaller, people managed to learn to do something, difficult to describe[212], that made only one muscle unit active.

So there are at least two classes of human action, automatic actions that are not under our voluntary control, and voluntary actions that feel as if under our control. We sometimes knock over a piece of

---

[212] But then so are whistling, riding a bicycle, and many other skills.

furniture because our body makes an involuntary twitch, perhaps when we have an illness like Parkinson's Disease. But we usually are quite clear that that is different from picking up a vase and hurling it across the room at an intruding rat. I don't think of the first of these as being "my" action, but rather something that is done to me or that happens to me. The second is something I do. The denial of free will is an attempt to say that all events in my life, even those that I think that *I really do*, are instead things that happen to me. Two questions, then: why should any one make what seems to be such an obviously false claim; and why should I be worried about it? Does it matter if things always happen to me rather than that I do them?

Ho-Hum[213]

I do many things "without thinking", even though it is certainly I who do them[214]. I will call such events in the biography of a human *actus hominis,* or HO. Many of them we share with the biography of other animals, and some (such as nutrition) even with plants. But none of this makes us think our freedom is limited.

No one would be worried if they thought, as indeed they should, that they had no control over their pupillary reflex, or even that a random muscle twitch had made them break a vase. The only part of human biography where worry about free will really arises is concerning truly human actions, *actus **hum**anis,* or HUMs. Indeed, the worry about whether we have free will is just that. It is a worry as to whether any of my actions are really mine, actions for which I am really responsible; or whether I am forced to do everything by some external or internal agency. Do I ever really HUM?

Let's grant that all my HOs are indeed caused to happen rather than my doing them voluntarily. Here are some examples of HOs as applied to Saltarella. The physiology of her digestion is a HO, not a HUM. She can't really control it. Her pupillary reflex is certainly

---

[213] A detailed account of HOs and HUMs is given in the next chapter. The Latin phrases come from medieval philosophy.
[214] D. Kahneman. *Thinking Fast and Slow.* 2012. Penguin Books.

a HO. So are all events in the nervous system viewed at the level of nerve cell activity (she does not directly command individual nerve cells), although the same events can be HUMs if viewed as the activity of the whole person when consciously thinking (she just decides to jump). Indeed if we think of the brain as being the vehicle of the mind, events at the level of the vehicle are always HOs, although the same events described at the level of conscious mental activity can be HUMs. In such a case we are talking about one and the same activity but from two different points of view. What about Saltarella's jump? Is it a HO or a HUM? She would certainly claim that she jumped, and was not forced to jump.

We won't consider HOs any further. Probably few of them are "free", caused by the exercise of "free will". But what about HUMs? These make us worry about the existence of free will, because they include actions that show us behaving as moral beings, as creatures who are good or bad, generous or selfish, cruel or kind. The reason that we worry about free will is precisely because we want to be able to claim that we exercise virtues and avoid vices. *We want to claim responsibility. We want to say that our lives do really include HUMs.* That is the issue of importance.

Philosophers have traditionally admitted that people certainly *feel* as if they are exercising free choice; but some philosophers as well as many scientists have claimed that the feeling of freedom is an illusion: in reality we are forced to do everything we do. They say that all the events in our lives are HOs, not really HUMs, as is implied in Maurice Evan Hare's limerick at the head of this chapter. Now, given how strong is our subjective feeling of freedom, of voluntary choice, why should anyone think that free will does not exist? Why should one claim that in a human life there is nothing but HOs? What is the relation between efficient or material causality and the nature of the will? For a start, what is "the will"?

It will not come as surprise at this point to find that I am going to suggest that there is no such *thing* as the will, and that here again we are being misled by the fact that our language talks about *the* will in a way that makes us think the noun refers to an object, a thing, a mechanism, a tool. We talk as if *I*, the mysterious real me, the *ego*, decides that I would like to do something, (that is I make a choice,) and I then use my will as a tool to give life to that decision. This makes it very surprising that there seems to be no equivalent word to the English *will* in Aristotle and Plato. But of course if Aristotle thought that an action was something a person performs without any intermediary tool, that would account for his language. We don't say, for example, that in order to run I have to use my runner; or that in order to speak French I have to use my Frenchspeaker. So why do I feel compelled to say that in order to act I have to use my will? Why don't I just say that I choose to act, or decide to act, or even just that I act, and never mention any kind of intermediary tool, spiritual or physical? Then instead of worrying about whether my will is free, I can just concentrate on asking questions about choosing, acting, and being morally responsible.

Could there be such an account? Let's see first some causes that might impede the freedom of the will, whatever the will may be. We have to consider at least the following[215].

Physical causality

The brain is made of nerve cells, and the nerve cells in turn of biochemical molecules and the molecules in turn of atoms; and the physical vehicle of the will, like the physical vehicle of the mind, must be the nervous system and its physical properties. Hence physical causality, the laws of physics and chemistry, must limit the freedom of the will. If the activities of the physical brain are limited by biochemical electrical functions, that must in some way limit what the will can do. It is this kind of concern that makes people excited to see the will in action by using fMRI to display the activity of the living brain. If certain groups of nerves fire, surely certain corresponding mental events, such as choosing what to do next, *must* be caused to happen. For example, it has been reported that when people try to respond rapidly, the response can be detected in the activity of the nervous system several seconds before the person is

---

[215] For a more complete treatment see G.Watson. 2003 *Free Will*. OUP.

aware of voluntarily initiating the response. (We will discuss this in more detail later.)

Genetic determinism

Many scientists are determinists as well as materialists. People understand the Story that scientists tell about human nature as saying that science expects, in the end, to be able to predict exactly what will happen in a human biography. They think, for example, that Dawkins's strong claims about the importance of genetic causality means that in the end a combination of genetics, chemistry and physics will let us predict exactly what choices a person will make, and what actions she will perform during her life. Even if we cannot make the predictions, nonetheless the actions must be biologically determined. That is not so. Paradoxically if it were not for genetic determinism we could not be free, because we could not develop into human persons. It is the biochemistry of DNA and cellular metabolism that guarantees that we have the kind of brain that is the vehicle of the mind and will. The genetics that gives us the kind of nervous system we have also gives us the ability to learn language (a mixture of HO and HUM) and hence the ability to perform HUMs.

In what sense if at all is genetic determinism relevant to the question of free will? There are some examples of the inheritance of single characters that do make a determinate difference to the biography of a person. For example whether someone will suffer from red-green colour blindness depends on the presence or absence of a gene on the Y-chromosome. About 15% of the male population have some degree of red-green colour blindness. Women don't suffer from this condition, which is indeed determined by a person's genetic makeup and cannot be avoided. Hence one's ability to make certain distinctions in the perception of colour is certainly genetically caused. Even here, however, there are different degrees of colour blindness. Not all those suffering from the condition have it equally severely. Most genetically determined characteristics are dependent on a large number of genes whose effects interact. Moreover in many cases the way in which the genes express themselves in the adult depend on events in uterine and early life, on diet (both of the child and of the mother during pregnancy), and on the environment in which the person matures, and hence on the person's own behaviour to some extent. Remember what we said, in Chapter 10, about the sources of Saltarella's abilities as an Olympian athlete.

Conditioning, learning, and training

Another common objection to free will is that people have been "conditioned" to do what they do. In their early life they were subjected to training, education, and even perhaps "conditioning". I put this last word in quotes because it was originally a technical term translated from Pavlov's Russian, not just another way of referring to learning[216]. Conditioning, to a psychologist, means that when some stimulus occurs, and the person responds correctly to that stimulus, she is rewarded. This increases the probability of making the same response the next time the stimulus occurs. Over a series of such co-occurrence of stimulus, response, and reward, the response comes to occur automatically without any "volition" by the learner. In human life the conditions leading to true "conditioning" are actually very rare. In particular learning language is not a matter of conditioning. The behaviourist psychologist B.F. Skinner made a heroic attempt to account for how language learning might be explained by conditioning, but it was destroyed from a logical and empirical point of view by Chomsky[217], and was never even slightly plausible. There is no reason to believe that much of linguistic behaviour is conditioned. Indeed there is very little about human nature that is plausibly described as being conditioned. When Saltarella learns how to perform the high jump, she is certainly not put in a position or a situation which is exactly replicated on every jump, nor does she receive direct rewards for performing exactly the same movements every time she jumps. Rather she practices jumping, and is trained to become ever better at performing a jump. This kind of training, athletics training, is diametrically opposed to conditioning. In most sports athletes do not learn to perform exactly the same movement

---

[216] In fact some translators use "conditional reflex" rather than "conditioned reflex", which gives a rather different feel to the semantics of the word.
[217] N. Chomsky. 1967. *A review of B.F.Skinner's Verbal Behavior.*

every time they perform their athletic feat, but rather to perform a set of goal-directed actions, which they may never have done before, to take account of the particular circumstances on a particular trial, on a particular day to achieve a particular goal (in all senses of the word). If Salterella tries to jump to a height that she has never before attempted, or modifies her style, by definition the situation has not previously been rewarded. The need for inventive rather than conditioned behaviour is even more obvious in team sports like rugby or soccer. Someone whose athletic moves were truly conditioned would be trivial to defeat.

Slightly different considerations arise in connection with the acquisition of complex skills such as car driving. Initially a person learning such a skill performs very selfconsciously a series of actions in response to different situations. As skill develops, these actions become linked into a series of "programs" which are run off when appropriate. For example, there is a chain of movements for changing gear, another for doing a start on a hill, and so on. Eventually these become more or less automatic, and when they become automatic they can be carried out without any deliberate choice. (Most experienced drivers occasionally realise that they have just driven for some time without being aware of the road or their own actions.) As this happens the environment triggers behaviour without any "willing" on the part of the driver. So a HUM becomes a HO and the behaviour in a sense becomes forced on the person doing it by the road. A HO may become compelling. Think of your automatic response to an emergency when driving in a country which violates your driving habits as to whether to drive on the left or right of the road: in an emergency you may automatically turn the wrong way. And as we will see, Mediaeval philosophers thought that virtue could become a habit.

Unconscious motivation, hypnosis, and social pressure

It is certainly true that we are all affected by motivational factors of which we may be more or less aware on some occasions. Whatever one thinks of the theories of Freud, he was certainly correct to claim that sometimes we do things for reasons or motives of which we are not aware. Indeed it seems likely that he was right to say that in some cases not merely are we not conscious of why we do things, but that we *cannot* discover what our true motives were without help: sometimes our motives are unconscious not just preconscious as Freud would have described it. There may be some actions which we are forced to do by unconscious motives, although such occasions are for the most part rather uncommon. Unconscious motivation, if it exists, would indeed imply a lack of free will, because by definition *unconscious* behaviour is not under voluntary control.

People sometimes invoke the phenomenon of hypnosis as an example of lack of free will owing to the possibility of post-hypnotic suggestion. This can't be very important, since few of us are ever subjected to hypnotic suggestion. But it does provide a case in which people think they have done something for one reason, have chosen to do it, when in fact they have been caused to do it, forced to do it, by the hypnotist. It therefore raises the possibility that there may be other cases in which we are mistaken about the origin and content of our motives and actions.

Hypnosis is a state of heightened suggestibility that can be induced in many people by a skilled hypnotist. The "victim" or "client" can be put into a sleep-like trance, although we know that the state is not physiologically the same as sleep. While in this state they can be given a suggestion that will change their behaviour once they are "awakened". For example, a hypnotist may say, "When you wake you will not remember that I have given you these instructions, but the next time someone comes into the room, you will take off your right shoe." The hypnotist then "wakes" the client, who does not remember anything that was said during the trance. However, the next time someone enters the room, the client does indeed take off his right shoe. If asked why he did that, he will tell a plausible story, such as, "I felt uncomfortable, as if I had a small stone in my shoe, so I took off my shoe to have a look."'.

It is commonly thought that in a hypnotic trance or a post-hypnotic state a person cannot be made to do things that are morally repulsive to him, but that is not quite true[218]. During World War II, an American soldier was hypnotised and told that when he opened his eyes he would see in front of him a Japanese soldier about to attack him, and that he would have to kill him to protect himself. An American officer stood in front of him and the hypnotised man was told to open his eyes. He attacked the officer, and had to be forcibly restrained from stabbing him. In another experiment people were shown that a glass vial contained acid that would attack the metal of copper coins. They were then hypnotised and told to throw the acid at an experimenter. Without them being aware of it the vial was switched for one containing harmless water. The hypnotised people did indeed throw the fluid at the experimenter. Furthermore, on one occasion there was a mistake made in the experimental protocol, and the hypnotised person actually threw acid at the experimenter, fortunately without the latter suffering harm.

So it appears that people, under hypnosis, can be made to do things that they would, in their normal state, consider morally wrong. And we might conclude that similar unconscious determinism might, in principle, act to influence or determine behaviour in other states of life[219].

At another level, we have plenty of evidence that people can be strongly affected in their behaviour by social pressure from their peers. Perhaps the most impressive demonstrations are those involving experiments on the effects of authority, in which people were persuaded to deliver extremely painful electric shocks to other

---

[218] T. X. Barber 1969. *Hypnosis: a Scientific Approach*. New York. Van Nostrand.

[219] It is worth pointing out a different interpretation of these experiments. Given the wartime setting of the first experiment, we could argue that if the hypnotic suggestion distorted the soldier's perception so that he saw he was being attacked, it was *not* morally repulsive for him to defend himself even to the extent of trying to kill his opponent. And in both experiments we might argue that although the participants in the experiment did not explicitly say so, even to themselves, they may have believed that in an experiment they would never be made to do so something evil, so that they could safely obey any instructions. Be that as it may, the *behaviour* was certainly dangerous in both cases,

people because they were told to do so by an experimenter[220]. If people cannot resist commands of this kind in a relatively benign experimental university setting, it is not hard to imagine that some people cannot resist commands to perform almost any act given the right sort of social pressure. Of course not all people will behave like this, because there are big cultural differences in how people respond to authority. The issue comes to the fore in war crimes trials.

Predictability

People say that because we are made of physical components, and hence subject to the laws of physics and chemistry, all our behaviour is in principle predictable, and hence is not free. And some people have appealed to the fact of quantum[221] indeterminacy to say that since physical events at the atomic and subatomic level are not predictable even in principle, there is room for us to affect the functioning of our brain in, as it were, the space left by quantum indeterminacy. It is our will, they say, that pushes events in such a way as to produce the behaviour that actually occurs.

Fortunately we need not discuss the relevance of quantum indeterminacy to free will, because the discussion of free will in terms of the predictability of human behaviour is obviously a mistake. If we worry about whether we are free because we want to claim responsibility for our actions, then an appeal to unpredictability must be wrong. I don't want my actions to be unpredictable, because if they are unpredictable I cannot be held responsible for them. Even I will not be able to foretell whether what I do will be good or bad. On the contrary, when I claim to be a good person I hope exactly that my actions *are* predictable. I claim that when confronted with a moral choice, whether to feed the starving woman and her child, whether to refuse to torture or bully someone, my actions *are* predictable. When I claim to be a good person I am exactly claiming that you can rely

---

[220] http://en.wikipedia.org/wiki/Milgram_experiment. Recent research has cast some doubt on the extent to which the experimental subjects really behaved as described.
[221] J. Eccles. 1948. *The Neurophysiology of Mind*. Oxford.

on me helping the poor and being kind, on all occasions when I have the opportunity; not that on each occasion a die is thrown by someone or something and its outcome determines whether I make the "good" or "bad" choice. To claim moral responsibility (which, I suggest, is why we worry at all about free will), is precisely to claim that you *can* predict my actions under almost all conditions.

Our Everyday Story of virtue seems to have a rather puritanical emphasis. The virtuous person is one who can overcome the terrific struggle required not to be bad. But again it is useful to look at what the mediaeval philosophers said. Aquinas for example, again following Aristotle, said that the really good person is the one who has practised being good so often that he or she chooses the good action even without thinking about it. Such a person has a *habit* of being good. When you think about it that makes sense. A friend who is trustworthy is one on whom we can rely to choose the right action on all occasions. She is strongly predictable in her moral choices precisely because she is free.

We can go even further and point out that logically a morally free action must be *more* predictable than a deterministic action. Suppose that someone's moral choices are really entirely determined by physical, psychological, and social causes in the everyday word. How could we ever predict what she will do? There is a potentially infinite set of causes that could influence the outcome. What is happening to the molecules in her cells? To whom has she spoken in the last hour or two? What did they talk about? What experiences have there been in her past life that have taught her, conditioned her, rewarded her, for making certain kinds of choice rather than others? What is her genetic make-up? In what state are her blood hormone levels? In what state are the ribosomes and mitochondria of her cells? How efficient today are the neuromuscular systems that she uses to perform her action? What did she have for breakfast? Is she upset by a quarrel with her mother? If her action is really determined, caused, by any and all of these things, they are so numerous that we could not possibly predict what she will do in any situation. An infinite number of dice are thrown every second, and on the sum of their infinite number of faces the action is determined. If quantum indeterminacy has to be considered, the situation is even worse. But to claim that she is making a free choice is precisely to claim that many or most of these can be discounted, that she will choose to perform a kind and good action. I don't claim absolute certainty, because there may be some special consideration acting on her today. But on the whole, there are only a few things I have to take into account to predict whether she will buy food for the starving woman. Has she been recently paid? Has she her purse with her? Is she known to be generous? Since there are are far fewer relevant causes if she is free than if her behaviour is strongly determined, the outcome must be more not less predictable! Free action is more predictable than determined action. But equally, this story shows that the whole notion of casting the discussion of free will in terms of predictability is mistaken. The fact that the fall of a roulette ball is unpredictable does not make it morally responsible for anything.

Philosophical Fatalism

Philosophical fatalism is the claim that what happens must have happened and there was no possibility that anything else might have happened. It is a very abstract way of looking at the problem and looking at the world. It does not even require an account of the causes that led to events. Rather it makes a logical point, that since we know what has actually happened, it does not make any sense to say that perhaps something else might have happened, but did not. Although much has been written about philosophical fatalism, and much of it is interesting and logically subtle, I am not going to talk about it in this chapter. That is because what I will propose in the end as an account of free will is a way of looking at freedom and responsibility that is immune to any kind of deterministic account. That being the case, it is also an account of responsibility that could be given even by somebody who believed in philosophical fatalism.

Freedom and Responsibility.

All these approaches ask whether human actions are absolutely "free" or constrained in some deterministic way: that is how discussions of free will are usually conducted. But if we look carefully at the notion of "freedom" in this context, there is something rather strange about it. Think of a simple action like raising your arm. Or ask whether Saltarella's action at the moment that she jumps was "free" or forced upon her.

Well, there were certainly strong constraints on what happened. Her movements involved strongly deterministic processes. Somehow she commanded her muscles to contract at the right moment, to the right extent, and in the right order. She certainly implicitly relied on the determinism of physics and biochemistry to make sure that her muscles contracted in the right way. But what about her *decision* to jump? Well, again, we don't want it to be free in the sense of there being no cause for it. If that were the case, then the jump presumably could occur at a random moment and in a random place! But we don't want her just to leap about suddenly at any old time and place like a drunken kangaroo. In saying that she freely chose to jump, we obviously don't want there to be *no* cause or reason for the jump, but rather one of a particular kind, namely that she decided to do it. We don't even want to say that by freedom we mean there is no physical cause. That can't be be right, because after all the moment for her takeoff is chosen in the light of the visual information array which enters her eyes, triggers responses, and activates the necessary pathways in the brain so that she sees that she is in exactly the right place to optimise the jump. So we don't want her to be free from the determinism of events in the nervous system.

The same would be true if we were not considering a physical action like a jump, but a moral action like giving money to a poor person. We don't want there to be *no* reason for this, no cause for it. If that were the case it could not be a virtuous action. For it to be a virtuous action on the part of the giver, we must believe that there is a cause, namely that the giver believes that he should support poor people with money, and is able to move his hand and arm in an appropriate way to find a wallet, take out the money, and to give it to the beggar.

So if asking whether one is "free" from physical determinism is the wrong question, what is the right question? We don't want to show that a person is free from constraints in the sense that a roulette ball is free to fall into any slot in the wheel, or that atoms are free to disintegrate at arbitrary moments if they are radioactive. The real worries about freedom are that they refer to whether and not we feel ourselves to be *responsible* for what we do. If we could just be certain that we were really responsible for our actions, our thoughts, and the events in our lives, we would be satisfied. We would not need a story about *the will*. We would not need to think in terms of being free. We would indeed be satisfied even if we knew that the events in our lives were determined in a strong sense, providing we could feel that we were responsible for them. But how could that be?

We have seen all sorts of reasons for thinking that behaviour and events in a human biography are causally determined in a strong sense, in ways that feel as if there is no room left for responsibility. Well, let's look at a case where I admit that I have been forced to do something, and ask whether there is any way in which even so I could take responsibility for what happened. The event might be somebody grabbing my wrist, forcing my hand into my pocket, closing the fingers on some money, dragging the hand out, and pushing the hand towards a beggar, then opening the fingers. But let's consider a slightly more fanciful Story.

Hypnosis and Responsibility: a fable.

You and I go to a theatre where someone is demonstrating hypnosis. I volunteer to go up on the stage and be hypnotised. The hypnotist puts me into a trance, and while I'm in the trance tells me that I will give money to the first shabbily dressed person I see in the street after I leave the theatre. Furthermore he imposes on me a post-hypnotic suggestion that I will not consciously remember those instructions. He then brings me out of the trance, and completes the demonstration. The hypnotist says to me as I leave the stage, "Do you remember any commands I gave you?" And I reply truthfully, "No, I don't remember anything that you said while I was hypnotised." At the end of the show you and I leave the theatre and after walking some distance through the streets, we are approached by a rather ragged looking figure. I take out my wallet and give him some money. You say to me, "Why did you do that?" I reply, "He looked very wretched, so I thought I would give him some money to get a square meal and buy some new clothes." You say, "Actually, the reason that you gave him the money was because while you were hypnotised you were given a suggestion that you would do so but you would not remember that you had been given the post-hypnotic suggestion." And then I say, (and this is the critical move to preserve my responsibility,) "Well, that may be so. Although I don't remember it, I am quite prepared to admit that I was forced to give him the money by the hypnotist. But *that is an action that I would like to have done, even if I had not been forced to do it. It is the kind of action that I approve of. It is an action that had I been free to do so, I would have chosen to do. The fact that I was forced to do it may be true, but I approve of it. I claim it as **my** action.*"

That is enough to take responsibility. It is neither more nor less than the way I use that phrase on other occasions in everyday life. I take responsibility for damage caused by my dog, or my small child, and ask a friend to treat the damage as if I had done it, and let me pay for it; so in this case I take responsibility for the events which were forced on me. Notice that I am not saying that I was free to perform the action. In fact I admit that I was forced to do it. But by passing judgment on the action I can in a very real sense make it mine. And by doing that I take responsibility as if I had been free to perform the action.

And that is all that is required. If I can be responsible for an action that I admit I was forced to do, then in principle I can take responsibility for any action. But the responsibility does not arise because I was "free" to do it and there was no cause that made me do it, nor because I had the use of a ghostly "will" as a tool, but because I am the kind of creature that has language with which I can pass judgement on events. Someone may say, "Aha! But then you were forced to make that judgement because of your past education! That's how you were conditioned by your family!" And I can reply, "I don't care: even if my education forced me to pass that judgement, when I think about it, it is the kind of judgement that I would like to have made even if I had been free, and not forced by my education, to do so. What a good education it was to give me such values!" Freedom is not necessarily a very useful concept: but the claim to responsibility gives me all that I require to be fully human. So that is enough. There may be no such thing as a free will, indeed no such *thing* as a will at all, but there can be responsibility none the less.

Freedom, Responsibility and Neuroscience

To end this chapter let's look at some neuroscience research that has mistakenly been thought to undermine the freedom of the will. Some years ago Libert conducted experiments which claimed to show that when someone responds to a stimulus the neural activity in the brain starts well before the conscious willing of the movement[222]. The following is from the New York Review of Books.

> [223]The subject watches a video screen where letters appear one by one. His task is to say what letter is on the screen when he freely decides to push a button. On the monitor, decision-making activity in his brain is visible as much as ten seconds before he is conscious of making the choice to push the button.
>
> "Neuroscience results are so dramatic, ... Eight to ten seconds before you feel you have made the choice, there are already conditions that are affecting your choice, determining the choice for you. What unconscious mechanisms are driving your decision? And is it still free will? For a decision to be freely made, you have to be aware of what went into the choice."

This is said to have implications for how "the will" works, because if the action starts before the conscious willing occurs, it cannot be that the action is caused by the will. Hence free will is an illusion. That is clearly nonsense. It begins with an unspoken assumption that there is a fundamental difference between the nature of the will and the nature of the motor physiology, i.e. it accepts the Cartesian assumption of the Ghost in the Machine, the GIM.

---

[222] B.Libet, 1983.
[223] http://aspire.princeton.edu/news/archive/neurophilosophy.xml

The required act is to press the button and report what was seen. Who (or what) is it that performs the act? Answer: it is the person, who is composed of a brain and body with which he or she performs acts including mental acts. The person agrees to take part in the experiment, that is to perform all appropriate necessary acts, and by so doing presets herself to respond to the stimulus. Now there are many ways in which one can respond to stimuli, and in many cases I do not know how I do it. That is because I have learnt skills, and in learning I have formed my brain and body into a system that can respond even if I am not aware of what I am doing. For example, I do not know how to make my bicycle balance, I simply decide to ride my bicycle and by so doing I switch on a whole repertoire of movements and controls that make use of neural mechanisms of which I am unconscious in order to carry out appropriate actions in response to inputs to the visual and vestibular (balance) systems. Many seconds after I have decided to ride my bicycle I may unconsciously make a movement to compensate for a bump or a pothole in the road. But I am clearly not forced to ride my bicycle or to avoid the bump. Other examples are very rapid responses (parry-riposte) when fencing, or the times when I drive a car when "asleep" for a significant period. The fact is that the decision to take part in the experiment and to "react to what appears on the screen" is prior to the stimulus. The participant sets

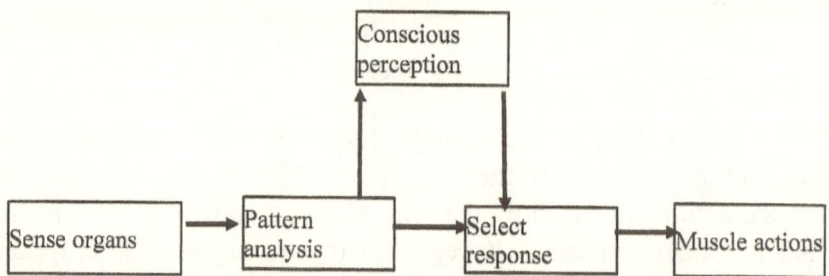

Figure 13.1 A neural information processing chart for the Libert experiments. For simplicity assume that it takes one time unit to traverse each arrow.

herself up to use all possible activities including those (comparable to blind sight) that are unconscious but allow a rapid response. There are lots of actions I "will" but do not deliberately and thoughtfully perform – balancing when cycling; rapid riposte in fencing, etc.. The initial electrical activity in the brain of which I am unconscious is still mine, not anyone else's, even if I am not conscious of it.

To fully understand this argument we need to know a bit more about the nature of information processing in the brain. When agreeing in the first place to do the task I (voluntarily) agree to do whatever I can to respond accurately to sensory signals, and that includes using "preconscious" "blind sight", or any other mechanisms as well. So if those are faster than conscious mechanisms, they will initiate a motor output before I turn on the consciously driven outputs. But all of them are done by "me" not by a mechanism that is a tool that I use, nor by a mechanism that uses "me". This also deals with the "Gunfighter Paradox." In Western movies the villain always draws first, but the hero, drawing in response to the villain's movement, shoots first. This is a cinematic convention, but is it realistic given what we know about neuroscience and voluntary action?

An informal experiment was once run in which people simulated the "fast draw" of the classic gunfighter film. Almost always the person who drew second fired first. It is faster to respond to the opponent drawing his gun than to be the one who draws first. The above analysis explains this. Assume there are two channels from the eyes to the muscles. One goes from the eye through neural pattern analysis pathways to conscious perception to a voluntary response. The other goes from the eye through pattern analysis direct to the muscles. I can program myself so that anything that is recognisable as a hand movement by my opponent triggers the second system. The one who draws first makes a voluntary decision to begin his movement. The hero does not have to make a voluntary reponse, but can preprogram himself in advance to let the his unconscious recognition of a handmovement by the villain trigger the unconscious pathway, which, having at least one less stage in the nervous system, will be faster. The person who draws first must necessarily make a voluntary and hence slower action.

The whole person is what responds, not "the will" and "the brain" separately. The voluntary action is to decide to take part in the experiment at all and hence to pre-program the entire response mode. A fast reflex unconscious response style is voluntarily selected and primed. The actual individual movement then is automatic, not voluntary.

This is another version of "taking responsibility for something I have to do". At the risk of being slightly misleading, I say as it were, "As soon as the sensory input reaches the part of the brain from which output can be generated, start the output. I will take responsibility for whatever you, oh my brain, subsequently tell me you have done." The allocation of responsibility is performed not at the moment of the response, but during the learning of the skill and the decision to use it whenever it may be needed. (One way to make sure that you never shoot anyone is to decide not to learn how to use a gun.). There is indeed no such thing as free will because there is no such thing as *a will*. But there *is* nonetheless moral responsibility, and you can claim it. What is its content, what counts as good or evil, is another question, and is for you and society to decide.

To end this chapter we can use it to clarify the relation between science and philosophy. Philosophical analysis does not deny the facts discovered by neuroscience or any other kind of science. But how to interpret those findings, how to relate one kind of Story to another, is not just, if at all, a matter of scientific method. Whether it makes sense to identify an electrical impulse in the brain with an act of will is not a matter for science except in forming an operational definition for planning research. But whether the operational definition is sensible, given the rest of our language, is not an empirical but a logical question. Stories, whether scientific or otherwise are reported in language and reflect how we think and talk about our lives; and to clarify the meaning of one kind of Story and relate it to a different kind of Story about the same facts is the province of philosophy.

Chapter 14

All About Souls

Animula vagula blandula,
Hospes comesque corporis[224].

> Emperor Hadrian, *Rome, 138 AD*

'Soul' has come to mean an immaterial immortal mind.

> Anthony Kenny, *The Metaphysics of Mind.*

The best picture of the human soul is the human body.

> Ludwig Wittgenstein, *Philosophical Investigations*

So we come to the last of the Fundamental Words, the word that in our culture is either used as a marker for human nature or rejected as an archaism no longer needed to characterize humans, the *soul*. It's probably the word most used in Everyday Stories when talking about what makes humans human, and also the word supposedly most threatened by modern scientific research on human nature. Looking back, although we have examined both the scientific account of human nature and the philosophical analysis of mind, will and life, we don't seem to have needed to speak of the *soul* very often. Yet in Everyday Stories, whether of a religious origin or merely a secular discussion that draws implicitly on ideas inherited by our culture, people still use the word frequently. Is there really a need for it? What does it really refer to?

---

[224] "Dear little fleeting pleasing soul/Guest and comrade of my body"

This will be a difficult chapter. But then, we are deep into the intuitions we have about the most profound aspects of human nature. There is little doubt that most people use *soul* to mean not just a Ghost in the Machine but the Ghost of a Person, a non-physical part of a human that is in some sense the "real person". It is a principle of life that involves one's identity and even for some offers the possibility of surviving death.

The word or its equivalent in other languages, *anima, psyche,* etc. is very old, indeed as old as Western philosophy itself. Whether we need it to describe human nature depends, of course, both upon what it means today, (since its meaning has changed down the centuries,) and also on how it relates to other kinds of knowledge, such as science. As Wittgenstein said, the meaning of a word is the use to which it is put, not the thing it refers to. I believe that there is an important use for the word *soul*. It is needed to emphasise a particular characteristic of human nature. But it may not be very similar to the current use in Everyday Stories.

At least as far back as the 4th century BC Greek philosophers used a word with a similar meaning. Today it appears in newspaper articles, in discussions of abortion, stem cell research, euthanasia, and in discussions of ethics. But you seldom see any attempt in popular discussions to say clearly what people mean by "soul" other than a vague feeling that it is a GIM. I hope by now the reader has come to see that a GIM is not a coherent notion. So let's look at the original meaning of the word *soul*, some of the ways it has been used down the centuries, and what we implicitly mean by it in Everyday Stories today. Then we can decide whether we still need it.

The English word *soul* comes from the Old English *sawol* and has been used to translate many words found in other languages, for example, $\psi\upsilon\chi\eta$ (psyche) in ancient Greek, *anima* in Latin, *napistu* in Hebrew, and *âme* or *ésprit* in French. In what follows I shall concentrate on the Greek and Latin roots of our tradition, because these are the most influential origins of today's Everyday Stories.

The word is used almost exclusively today in discussions of the truth or content of religion, but its original use was not primarily religious.

It was part of the philosophical analysis of the nature of humans and particularly of the difference between living and non-living things. Its restriction to religious discourse is a relatively modern development, and one that is not necessarily helpful or enlightening.

One thing we can say with confidence is that in English having a soul has always been thought to be what makes people alive. Furthermore for many people the nature of the mind, the will and consciousness are all closely linked to the idea of the soul. So having a soul makes us alive. But the previous chapters have I hope made a strong case that the origins and causes of life are chemistry and physics. So what room, if any, is there for the soul to be the cause of life in general and human life in particular? Was there a missing cause in our earlier accounts, one to which the word *soul* points?

Sources of confusion.

To understand words we have both to understand the languages from which they come and also how the people who spoke them lived. As we saw in Chapter 6, often to translate a word from its original language into modern English makes it look like a *thing*, but its original meaning was different. Take the word *being* for example. A *being* may mean a person (as in "a human being"); but it may also be shorthand for a philosopher's technical phrase such as "that which exists". The word we translate by *substance* to certain classical Greek philosophers was almost like what we would mean by quarks, that is, what everything is fundamentally made of, while today it is more or less synonymous with "a chemical". *Intelligere,* from which we get the word *intelligence*, in classical Latin meant "to understand", but in the Middle Ages meant also "to think about", which has a very different emphasis. (I may well think about something without being able to understand it!)

There are many such problems of translation, as we saw in earlier chapters. As late as Descartes most philosophy was written in Latin. The writers of the Middle Ages such as Aquinas and Ockham all wrote in Latin. Some of what they read of the works of Greek philosophers such as Plato and Aristotle had been translated into Latin by the time they came to know the works, and some of them read the original Greek. What comes down to us may be an English translation of a Latin translation of a Greek philosopher's work, sometimes preserved in an Arabic translation in the so-called "Dark Ages".

This causes difficulties. For example, remember that Greek had a definite article corresponding to our word "the", but did not have an indefinite article corresponding to "a" or "an". (They did have a word for the number "1" and a word τις (tis) that is usually translated as "a certain..." but could sometimes stand for the indefinite article.) Latin has neither definite nor indefinite articles. So if we were to translate the Greek 'η ψυχη which means roughly "the soul" (the psyche), into Latin we would use the word *anima,* or perhaps *mens*, although the latter is less likely. But if later someone translated the Latin *anima* into English, he would not know whether to say "soul", "the soul", or "a soul" except by understanding the original context, the way in which the original Latin author had been thinking.

Suppose the original Greek were, 'ανΘροπos ψηχην 'εχει, which means roughly "a human has a soul", and someone translated it into Latin in the 13th century as *homo animam habet*. Then suppose that in the 19th century someone translates the Latin into English. The English translation might come out as any of the following:

1. Man has a soul
2. The man has a soul
3. Man has the soul
4. The man has the soul
5. The man has soul
6. Man has soul
7. A man has a soul

All of which have subtly different meanings and some of which don't seem to mean anything!

Let's look at translations 1, 3, and 5. In modern English I can imagine these sentences meaning quite different things, including,

- Man has a non-physical component.

- Man (but not, say, an armadillo or an aubergine) is the creature that has a non-physical component.
- Man has a quality or property (but not a component or part) that other creatures don't have, but it is not necessarily either physical or non-physical.

But are any of these statements really what those who originally developed the ideas would have meant? Does the way we use the words today agree with their original meanings, and if not, to what extent have we lost important overtones of meaning and use? We may have even clarified the meaning compared with what it meant in the past. Aristotle discussed about a dozen meanings of the word for *soul* that had been used by philosophers before him, decided that they all were wrong, and then put forward his own view. Aquinas followed Aristotle closely, but moved from Greek into Latin; and neither of them meant what Plato or Descartes meant.

Our modern Everyday Story is more like that of Descartes than anyone else, since it is certainly true, as Kenny observes, that

> 'Soul' has come to mean an immaterial immortal mind[225].

But he goes on to observe that,

> For pre-descartes Aristotelians, it was merely the principle of life in living things.

Don't be surprised if what follows does not sound like what *anybody* commonly says these days; for I think that in some respects what was said hundreds of years ago makes more sense than the Everyday modern view. Accepting Kenny's comment on Everyday use of the word, let's start with the most common modern way to approach the notion of *soul* in a Christian Everyday Story.

---

[225] A. Kenny, 1986. *The Metaphysics of Mind*. Oxford: Oxford University Press. p. 18.

Neville Moray

An Everyday Religious Story about the Soul

By "Everyday" in this case I mean the kind of Story that is told in newspapers or radio broadcasts by non-philosophers, or in religious sermons. For some reason the speaker has come to believe in God and more particularly in Jesus. She believes that God has told her that Man has a body and a soul. The first is material; the second is "non-material" or "spiritual". Christ died to help her "save her soul" so that when her body dies, the soul is freed from the body and goes to Heaven to "live" forever with God. Having a soul is what makes a person human and alive. My soul is the "real me" and is created specially by God when conception occurs. My soul is in my body rather like a captain is in a ship, guiding and controlling my bodily actions. Since each person's soul is created by God at conception, and since the human soul is "the real me", then an embryo is a human being. You are not allowed to kill humans, so you must not kill an embryo. If stem cell research means the death of the embryo, neither stem cell research nor abortion is permitted, nor is it permitted to use the blastula, the ball of cells that appears after the first few divisions of the fertilized ovum, as a kind of stem cell factory.

Few Religious believers would try to prove all this. It is mainly a set of philosophical ideas to make sense of what they believe has been revealed by God. For Descartes it was an attempt to give some meaning to the word "I", that is, to the idea of personal identity. The notion of the *soul* as a spiritual, immaterial component connected to a physical (material) body was appropriated as a religious notion to make sense of beliefs about life, death, immortality, free will, and so on. There are other Religious accounts of the soul, but I think what I have written is a fair summary of how a fundamentalist Christian would talk or write today.

With the development of stem cell research and therapy, cloning, neuroscience, genetic engineering and the increasing use of abortion, the soul remains a central concept when the ethics of human development from a fertilized egg is discussed. To make it easier to see how the idea of a soul relates to the idea of a developing human, Figure 14.1 provides a summary of the first months of human life. A religious account would be something like the following.

After copulation God guides a spermatozoon to an ovum and when they join he creates the soul of the new person. The soul guides the development of the body and causes it to become a blastula, then an embryo, then a foetus, then a baby, then a child, then an adult. At death God takes the soul to himself for ever, freed from the limitations, temptations and constraints of the body. The body decays. During life the soul causes the person to develop and interact physically, intellectually and emotionally with the world and all that is in it. (Many people say that humans are the only creatures that have souls.). If we agree that we are not allowed to kill human beings, the question in abortion is whether a blastula, embryo or foetus is a human being. Since God has given a fertilized ovum a soul, it is human and we should not kill it, and that includes dismembering it to make other cells.

But why do we think there is such a things as a soul in this sense? Do we ever encounter one, even our own? How could we? In everyday life we do not see souls: we see living people. If we think that the reason they are alive is that they have a ghost in their machine, a GIM, then it must be because we first decide they are are alive – that is what we can actually be sure of. If I look inside myself by introspection, I certainly don't see a GIM, nor am I aware that I am really a ghost. Far from it: given all the aches and pains that afflict me every day, what I am aware of is being a body. I never encounter *me*: all I know is that I perceive, feel, think, decide, and act. Any reference to a GIM must be a deduction from such observations. But why do I need to make such a deduction? When someone dies, why should I say that his soul has left him? I see no such thing happen. What I see is that he stops behaving in the way that an integrated living person behaves, as we saw in Chapter 9. At first some parts of him continue to behave as independent organs, but eventually they too stop behaving in a coherent way (showing their normal functions) and fall apart into their chemical components. And as we saw when talking about anaesthesia, a person is sometimes, when unconscious, not even the best judge of whether he is alive or not, whether his soul is present or not. So why do we need the notion of soul at all?

A scientific account of development[226]

Here is an alternative, non-religious account of the biography of a human being, which still finds the notion of a soul useful. Spermatozoa and ova exist in men and women respectively, as living cells supported by the biological environments of the bodies of their respective owners. Each has half the chromosomes of an adult. Left alone, neither will develop into any other kind of organism, nor divide to make a more complex collection of cells. Following copulation, the sperm enters the ovum, and a new animal, a fertilised human ovum or *zygote*, is formed and begins to develop. Now that the sperm and ovum have become a single organism the latter behaves in a different way from the previously isolated spermatozoon and ovum, and begins to divide to make a collection of 2, 4, 8, 16 etc., cells, and so turns into the blastula, the ball of cells that is the form a human being has when it is at the 16-cell stage of development. (It is of course a human blastula, not the blastula of an armadillo or an aubergine.) Up to this stage all the cells can turn into more or less any kind of cells and hence make almost any body part. In effect they are all stem cells. Now constraints appear, and driven by genetic and biochemical mechanisms the geometry of the organism begins to change. Cells of the blastula differentiate into nerve cells, digestive tract cells, muscle cells, etc., and the shape of the whole organism becomes more linear and bilaterally symmetrical. In a few weeks the nervous system begins to form, the primitive heart begins to beat, and a viable foetus develops from the embryo. The latter in turn becomes a baby, a child, and an adult. As language appears and develops there is a final but most important differentiation, that mediaeval philosophers called the difference between the ability to perform *actus*[227] *hominis* and *actus humanis*. *Actus humanis*, truly fully human acts, require human language, but everything from fertilisation onwards is an *actus hominis*, that is an event in the life of a human being appropriate to its then stage of development. As in

---

[226] For an excellent introduction see Royal Institution Christmas Lectures, 2013. BBC television.
[227] *Actus* in Latin is both singular and plural, "act" and "acts".

Science, Cells And Souls

the previous chapter I'll refer to these different kinds of behaviour as HO (short for *actus hominis*) and HUM (short for *actus humanis*).

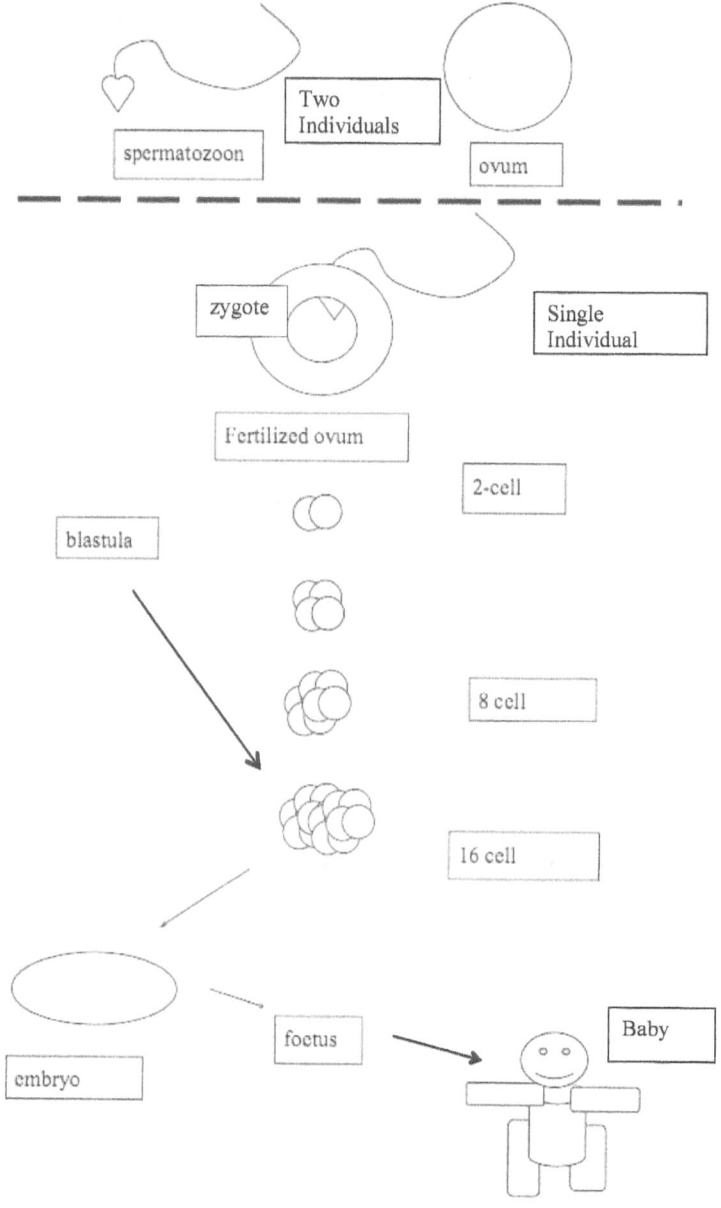

Figure 14.1 The Development of a Human

At every stage of development after the formation of the zygote the creature is alive in the way that a *homo* is alive at that point in its developmental sequence. That is what it means to say it has a "soul". It is never an armadillo, never an aubergine, an amoeba, or anything else, but it *is* alive in the way a typical *homo* is. (Remember in Latin the word *homo* means "a man" in the sense "a member of humanity", whereas *vir* means a man as distinct from a woman, *mulier*.) The developing organism may not yet have reached a stage at which it can perform HUMs, truly human acts (as distinct from acts that any other animal, or in some cases any other living thing can do,) but since its biography consists of HOs, it is a developing human and should not be killed. It is not just a collection of biological bits and pieces, but an organism: and the fact that we need to talk about it as an integrated whole, not as a collection of parts, is what we mean by its having a soul.

We have arrived at the same conclusion as the fundamentalist Religious person, but without any need to postulate the existence of a non-physical soul.

Ho, Hum - But Not So Boring

Let's reprise a bit of the previous chapter. There are many things that I do "without thinking", even though it is certainly I who do them[228]. These are all HOs, events in the biography of a human, although many of them are shared with the biographies of other animals, and some (such as nutrition) even with plants. Indeed even some actions that initially require care and attention may become, with much practice, so automatic as to be correctly classed as HOs. Think of the difference between the deliberate, conscious, movements required to control a bicycle or car when first learning to ride or drive, and the fluent totally thoughtless skill of the expert. There are however other events that are uniquely human, and can occur only because we have the kind of language that humans have. They are unique in a special way; they involve voluntary, self-conscious action. An

---

[228] See D. Kahneman. 2012. *Thinking Fast and Slow,* for a modern treatment.

example would be someone who says, as the *Titanic* sinks, "I will give you the last place in the lifeboat because I love you, and I think it is more important for you to survive than for me. I will take my chance by swimming." This act of self-sacrifice is a HUM, a truly *human* act, not just an animal act.

HOs then are often shared with other animals, appearing as nutrition, reflex behaviour, etc. In sexual behaviour, very roughly, the decision to make love to someone *can* be, although it need not always be, a HUM, but ejaculation is a HO. But even though HOs are "merely biological", they do differ from species to species. I suppose we could say of our dogs and cats that they perform *"actus canis"*, *"actus felis"*, as we do *actus hominis*. And the behaviour of cells in the development from the fertilised ovum to the baby are also HOs, not acts in the biography of some other species. Hence development, *after fertilisation* is a sequence of HOs, although prior to fertilisation there are only the separate *actus spermatozoonis* and *actus ovi,* events in the biography of a spermatozoon and events in the biography of an ovum, not events in the biography of a member of humankind. Before the combination of the spermatozoon and ovum to make a zygote there are two individuals, two souls. Afterwards there is only one individual and hence one soul.

To claim the central importance of language in the ability to perform HUMs may seem to be a claim that a pre-verbal child is not "fully human"; and I think some mediaeval philosophers would indeed have said that. It is interesting that some of them said that children could not commit sins until about the age of seven to ten, (which is roughly when language is sufficiently developed for children to be aware of "right" and "wrong") or even later. I don't want to go further into this issue, but I note it, since it would make a difference in the abortion debate if one said that you must not kill humans who can perform HUMs but you can kill animals that can only perform HOs.

To summarise, for people who think in this way, if you have a creature that performs what you can identify as *actus humanis* you *certainly* have a human. If you have a creature that performs what you can identify as *actus hominis* you certainly don't have a creature that is anything other than a man or woman, (not an armadillo or an aubergine,) but you might have to wait several years until language developed to be certain that you have a truly *human* creature.

Souls old and new.

How is all that relevant to the idea of a soul? It is relevant because the original idea of a soul, for Aristotle, was very closely related to how something is alive. The two most influential ancient philosophers of the soul were Plato and Aristotle and it is from them that our Everyday Stories about souls are derived. Their work was known to the Romans, but much was lost to the Christian West after the collapse of the Roman Empire. Later, in the 11th – 13th centuries, Greek thought was rediscovered, preserved in Islamic libraries. By that time Christian theology had developed, and philosophers were discussing the nature of God, how humans could exist after death, how to make sense of the final judgement and resurrection (the "2nd Coming"), and so on. Some of these questions were theological, but others were more generally philosophical and were dealt with in a way that closely resembled the classical Greek discussions. In particular the group of philosophers called the Schoolmen, including Duns Scotus, Peter Abelard, William of Ockham, Albertus Magnus, and Thomas Aquinas made use of the rediscovered classical works. Some went along with Plato, the idea that the soul was the immaterial "form" of the body and was the "real person", of which the body was little more than an imperfect shadow. But others followed Aristotle and developed a different view. In the context of the present book it is interesting to note that of the ancient philosophers Aristotle was by far the most interested in natural science.

Here are some translations of what Aristotle said about "soul".

1. "Having soul is to have life."
2. "A soul is made up of material that is capable of autonomous change."
3. "A soul is not what causes change, it is not "the subtlest and most incorporeal of bodies". It is not made up of "elements" (for Aristotle earth, air, fire and water; for us hydrogen, oxygen, nitrogen, etc.).

4. "A soul is what a thing is made of in the sense of a formula describing what a thing really is."
5. "If you cut a plant or animal into segments the segments have the same kind of soul (though not the same soul)."

He said that plants have *nutritive* or *vegetative* souls, animals have *sensitive* souls, and humans have *rational* souls. Note that from (3) Aristotle did *not* think that the soul was an immaterial "ghost in the machine". From (4) he seems to have thought that a soul was more like a logical category to be used for description, rather than a part or component of the person. And in (5) he presents what we could see as a discussion of cell division. For Aristotle there is no problem about an amoeba that divides. After the division each amoeba has an amoeba-type soul, although neither is the same as the soul of the original amoeba before it divided[229]. As regards the developing human, he would probably say that the initial cell divisions lead to cells which each have their own soul, since they seem linked by physico-mechanical forces rather than being an integrated unity. As organs like kidneys, the heart, etc. develop and begin to function as autonomous organs, he might want to say that each was alive in its own way, and perhaps that a kidney, viewed as a coherent entity not just as a collection of individual cells, would have a "kidney soul[230]. After all, if being alive is the same as having a soul (from (1) above), then the individual organ must be alive in its own way even when it is not connected to the rest of the body – that is why organ transplants are possible. A kidney is alive in a different way from a pancreas (or it would not matter which you used for a transplant). Finally when all the organs function as a single unified individual there is only a single soul, that is a single person.

---

[229] It is amusing to imagine how amoebae would deal with the problem of names if they could talk. After a division the two new amoebae could not be the same one "person" they were before, since they are identical. I suppose they would have to have a Scandinavian kind of naming. "Hello, who are you?" "I used to be Fred. Now I am Fredson. And that person over there also was Fred and is Fredson."

[230] For a fascinating Fictional Story treatment of such ideas, see D.Lessing, *The Making of the Representative for Planet 8.* London. Flamingo Press. 1994. especially pages 86 et. seq. in that edition.

So a soul is what makes a living thing have its properties, but a soul is not an immaterial spirit trapped in the body, nor merely a lump of matter. In fact in Chapter 7 we implied that to say that something has a soul is to draw attention to the fact that it has a life as a coherent integrated entity, not just as a collection of parts. To have a soul is to be an identifiable individual. In this way of talking a soul is a description, a categorisation, not a causal explanation; and in this role it is useful to emphasise the difference between a collection of parts and a real, live, individual. It provides a way of turning a reductionist account of a body as a collection of molecules into a Story about a person.

Let us follow this line of thought further and see how Aquinas and Aristotle might talk about the soul causing things to happen in a living creature taken as a whole, perhaps the high-jumper Saltarella. Is the soul a difference in the way things are made, or just in the way we talk about them, an ontological or just a logical difference? Surely there's a problem about how a soul can "cause" properties of the person, and make events happen in his or her biography. How can the soul be the efficient or material cause of life in the body? We clearly don't mean a cause in the sense of DNA or physics. So how did the classical philosophers make souls work for them?

Plants, Aristotle said, have *vegetative* or *nutritive* souls; that is, they are alive in the sense that they take in nourishment and metabolise it to obtain their energy. Animals have those properties, and additional properties such as being sensitive to stimulation by light, touch, sound, etc., and also the ability to move and act. They are said to have *sensitive* souls. Humans have *rational* souls, because in addition to the kind of sensitive life they share with other animals they can reason, and the powers of the human soul were said traditionally to be memory, understanding, intellect, and will. For Aquinas[231] many

---

[231] Thomas Aquinas was both a theologian and a philosopher of prodigious productivity. He may have written more than any other philosopher. His work as it has come down to us comprises between 8 million and 10 million words, which suggests that he wrote something approaching 1000 words a day for 50 years! A tradition says that he would dictate to several secretaries simultaneously, although as a student he was known to his fellow students as the "dumb ox".

animals, even those with only sensitive souls, also have will, (for example a sheep may choose one field of grass rather than another,) but only humans can make morally responsible free choices. When a person dies, Aquinas believed (for religious reasons) that his soul does not cease to exist, but far from being liberated from the bonds of the flesh, the soul after death is not a person at all, and cannot perform any human acts (HUMs). Only embodied souls can perform HUMs, just as only ensouled bodies are alive. So when a person begins life as a fertilised ovum, and develops into a foetus and then into an adult, it's true that the soul is the cause of his or her life, because to be alive is to have a soul. But what does that mean?

As we saw in earlier chapters the reason you know that someone has a soul is because you know she is alive, not the other way round. That is why it is sometimes difficult to decide whether someone has died. If we really had direct evidence for souls there would be no puzzle: soul's present – person's alive: soul's gone – person's dead. But actually it is the other way round. Person's alive - soul must be present: person's dead – soul's gone. It begins to look as if "having a soul" is, for Aristotle and Aquinas, almost a behavioural concept. This also raises the question of how to talk about cases of multiple personality, where different people seem to share a body at different times. Presumably Descartes woud have to say that several ghosts were sharing a single machine, since each of the personalities might say *cogito ergo sum* on its own. If having a soul is a way of referring to the existence of an integrated person, then perhaps we would have to say that each of the personalities signalled the presence of a different person, and hence a different soul. Each personality has its own habits, its own memories, its own intentions. Since it is identifiable as an individual, Aristotle would I think say that the appearance, the occurrence, of each personality indicated the presence of a different person, and hence of a different way of being alive, and hence having a different soul. If we look again at items 4. and 5. in the list of what Aristotle said about having a soul, he would not be troubled if a change in the physical structure caused a new soul to be present, and that is what we see in a case of multiple personality. The brain functions differently for each personality. If a soul is not a *thing*, but a description, why should there not be more than one description associated with a single body?

Remember what we said about causes in Chapter 2. Aristotle and Aquinas distinguished four kinds of cause.

1. The *efficient* cause (E), namely what you have to do to make something happen.
2. The *material* cause (M), namely the underlying physical properties of the world that enable it to happen.
3. The *final* cause (F), namely the purpose for which it happened.
4. The *formal* cause (Φ), namely the set of properties that make us logically assign it to being a particular kind of thing.

Although most people in the Everyday Stories of our time seem to think that souls are material causes, Aristotle and Aquinas said that the soul is the *formal* cause, (Φ), of life in a living creature. To make these distinctions clearer, let us look at two examples.

What causes an animal to be a living creature?

1. The animal is alive because it catches and eats prey (E).
2. The animal is alive because it has a particular DNA and a particular kind of biochemistry and metabolism (M).
3. The animal is alive because it will be able to fill a particular ecological niche (F).
4. The animal is alive because when you examine it you find that its biography contains the kind of events that make you want to classify it as belonging to the class of things that are alive ("a formula that describes what it really is" as Aristotle put it) (Φ).

And now for what causes a blastula to be human.

5. The blastula is human because a spermatozoon and an ovum united following copulation (E).
6. The blastula is human because its cells have a certain kind of DNA and a metabolism that supports its cell division appropriately (M).
7. The blastula is human because it will eventually make a human adult (F).

8. The blastula is human because it is behaving (growing, dividing, cells migrating, etc.,) in just the way that a human blastula is expected to behave at that stage in its development (Φ).

In modern discourse we are used to thinking that only Material and Efficient causes count as "real" causes. (F) is what is today called "teleology", or goal-orientation, or "purpose", and is usually disallowed as a scientific account of cause. (There are some cases in cybernetics where it may be allowed, but they are special. Remember that Wiener, the founder of cybernetics, thought of it as the science of purposeful systems.) (Φ) *is what a soul is* according to Aristotle and Aquinas. It is a logical, not a "material" or physical cause. Perhaps a clearer example of a formal cause would be, "What causes a person to be an immigrant?" Answer: "He or she is identified, and put into the category, of "a person from another country taking up residence in my country". That does indeed cause him or her to be an immigrant, but it is a Formal, not a Material or Efficient cause. On the other hand, there does not seem anything peculiar about describing a logical classification as a cause in this case. By contrast, there is something very strange in the Everyday Story that says that a spirit is a Material cause!

The way in which Aristotle and Aquinas talk of the soul as being the *formal cause* of a creature being a human seems perfectly reasonable, and indeed useful. It is not a religious claim. It is a claim about what makes a person *count as* being human. Whether one believes in a religious account of human nature, or that something continues to exist after the body dies are quite separate questions from whether a person has a soul. For Aristotle and Aquinas a person obviously has a human soul because otherwise we would not be able to identify her as a living human. The notion emphasises the fact that a living person is identifiable as an integrated individual, not just a collection of parts. But the soul is not the "real me". In fact Aquinas explicitly says that, *"Anima mea non est ego."*, *"My soul is not me."*[232] How different from Descartes!

---

[232] Aquinas's commentary on I Corinthians 15 in P. Geach, *God and the Soul*. P.22 . Routledge and Kegan Paul. 1969. London

This way of thinking opens the way to a non-dualistic approach to human nature. For example, in the Aristotelian-Aquinas framework, if I wanted to argue in favour of stem cell research or in favour of abortion, I could argue like this:

> "When I look at a blastula I do not find enough properties to classify it certainly as being fully human. Certainly it cannot perform any HUMS. I don't even think it performs any HOs. It does perform *actus animalis,* since it is alive in the way that many animals are alive when they are developing, but it hardly functions as an individual, more like a loosely connected group of individual cells stuck together by the chemical properties of cell membranes. Over all, it is not, for me, convincingly classified among things-that-are-definitely-human, even though it certainly came from a fertilised human egg. So for me it is not yet "ensouled" in a human way. Its soul is not (yet) a human soul. If left to itself and if development continues normally it will later become human and have a human soul (behave in a way that makes it convincingly classifiable as human). But at present, for me, it is not behaving fully like a human, so it does not have a human soul. It can be killed or used for stem cells without worrying about killing a human because it does not have a human soul in the (Φ) sense."

For someone who wants to argue the opposite the way to do so should be obvious, and "is left as an exercise for the reader".

This way of talking includes the notion of a soul, but no religious aspect. It is not common to hear this kind of discussion today, but it seems reasonable. Notice that it sometimes leaves you uncertain as to whether the creature is or is not human. The concept of a soul as the "form" or "formal cause" of a being, is not a religious issue, but a philosophical concept to account for the manifest differences between living and non-living ("animate" and "inanimate") entities we find in the world, between just a collection of chemicals, cells or organs on the one hand and a functioning organism on the other.

Finally let's put the idea of a soul in a general framework of biology. An amoeba, or a bacterium, or other single celled organism (a monocyte), has a "unicellular" form (Φ) that means that it behaves as you would expect a monocyte to behave. When it divides we have two monocytes each of which is alive as is typical of a monocyte, and hence we have two new (Φ)s; two new souls have been created by cellular division. This is only a problem if we slip into thinking of a soul as a component of an organism, but not if we think of it as a logical classification. Why should a description not change when behaviour changes? Obviously to describe two individuals is different from decribing a single one which was their precursor. A blastula is an individual made up of a small group of cells held together by physico-chemical constraints (surface tension, electron bonds, chemical concentration gradients) all of which make (M) a loose individual. Each cell has its own life, and the thing-as-a-whole has a life-as-a-whole. As development proceeds, the biography becomes increasingly better described as that of a single integrated complex organism made up of parts, and the parts become unable to survive without being integrated into the thing-as-a-whole. So its form changes, and its soul changes from merely a sensitive soul typical of any animal to a sensitive soul specialised in the way that a human life is specialised, (which of course includes the ability later to develop language). From being pluripotent stem cells the cells become specialised and individuated in function.

The kind of soul that is present changes as development proceeds. As the foetus becomes more and more "typically human" we become more and more willing to say, "Yes, it has a human soul. It is definitely performing HOs, even if not HUMs." Eventually after it is born and becomes linguistically fluent we can certainly observe HUMs in its biography, and any doubt is removed: it definitely has a human soul. (Think how worried people are if their children are late in learning to talk.) We can talk about evolution: the metabolism (M), that made it behave like a particular animal and have a particular (Φ), is stressed by the environment (E), so that the biography of the animal changes. Hence the way-in-which-it-is-alive changes, the logical classification of its biography changes, its (Φ) changes, and it comes to have a different kind of soul, that is, to be a different kind of animal, alive in a different way. Kipling described this process as a fable in the short story called *The Ship that Found Herself*[233].

We can even throw light on the question of whether we can make a living organism from non-living materials. If we put together appropriate chemicals in a certain way they will interact appropriately, and the resulting compound will show new behaviour that was not present before. In principle there is no reason why that behaviour cannot come to resemble the behaviour seen in the chemical compounds found in living organisms such as viruses, bacteria and other monocytes. If the behaviours are sufficiently similar it makes sense to classify them as the same kind of thing (Φ), and hence to say they are "living". Have we then created a soul? "Yes, of course." would say both Aristotle and Aquinas.

Conclusion

Think again about the fact that we can use an Atomic Force Microscope to watch individual molecules perform their individual metabolic tasks. As Hoffmann[234] describes them we might almost think that the molecules are themselves autonomously alive. His "molecular robots" look like insects if we do not realise at what magnification we are seeing them as they go about their tasks of moving chemicals and energy from one part of a cell to another. But it is the organism that is alive, not the molecules. And the use of the word *soul* gives us a marker to assert the difference between the reductionist Story of physical chemistry and the alternative Story of the entire living organism. When we talk abut a creature having a soul we assert that there is an appropriate Story about it at the level of the whole integrated organism playing out its natural history. No ghosts in the machine, but much more than molecular chemistry; even perhaps social work rather than physics!

So, do souls exist? Do we need the word to discuss human nature? Yes and no. Souls are not *things*. There is no reason to think that we

---

[233] R. Kipling. 1898. *The Day's Work*. Doubleday.
[234] P. Hoffmann, 2012. *Life's Ratchet*. Basic Books.

are bodies inhabited by ghosts. But it *is* important to realise that we are more than a collection of chemicals. Just as the Electrician was forced to appreciate that his collection of electrical components was more than physical components because it was an advertisement, so a living person, indeed any living creature, is more than a collection of molecules. It has properties that make it a plant, an animal, or a human. The whole *is* more than the sum of the parts, and *that* is what having a soul means. Nothing-buttery will not do when we talk about humans. Saltarella does indeed have the soul of an athlete.

That may seem trivial. But to be an athlete is to be human, and to be human is to have all those characteristics that we know from our own lives and those of our friends to be of the utmost importance. In Everyday stories people often say that our soul is what makes us have spiritual and transcendant qualities. We can love one another. We can sacrifice ourselves (although not our selves) for one another. We can appreciate beauty and we can be brave, charitable, friendly and morally responsible. In short we can make HUMs of our lives. As we saw much earlier when discussing philosophy there can be things that are made of matter but whose properties cannot be reduced to physics and chemistry. It is indeed true that to have a human soul allows us to have spiritual and transcendental aspects to our lives, not because the soul is a spirit, a ghost, in the neural machine, but because to be alive in a human way (which is what it means to have a human soul) is to be able to perform HUMs, and these cannot be reduced to a description in terms of physics and chemistry. Without the vehicle of a body and its nervous system a human has no soul, no life. Thanks to being alive, an ensouled integrated individual, all human nature is available, including particularly the ability to express in language our intentions, experiences, desires and satisfactions. HUMS are the properties of humans, of persons, not merely of brains and muscles. They can be asserted only of a person, not of evoked potentials in the brain or fMRI records. These human characteristics are spiritual, indeed transcendant, not because they take place in a spirit, a ghost in the neural machine, but because like the advertisement, they cannot be described in a reductionist account using only physics and chemistry wthout mentioning the person. Our language is a gift from our nervous system and DNA. Our spiritual characteristics are a gift from our language, not from our ghost, which does not exist.

We are back to the discussion of *Stories* with which we began this book. Living entities can be described at many levels and by many kinds of Stories. Humans can be described in terms of chemistry and physics at the level of the molecules that make up their bodies; or they can be described by their personal biographies, accounts of what they are doing, their experiences, intentions or plans. Moreover, these different Stories can apply at one and the same time. But if we look at the events at a molecular level we cannot see what the intention of the person is; and if we look at the expressed intention there is nothing to say what particular molecules in which particular cells are involved. Indeed we don't need to know that in order to know their intentions: they can tell us. Because there are so many ways to perform any HUM, so many ways to be kind, so many ways to be cruel, there is an uncoupling of the Stories at different levels, and yet they all apply to the same person. That is what it means to claim that they have a soul. When this multiplicity of Stories ceases, when only some Stories but not others are applicable, the person is no longer alive, and so we say that they no longer have a soul, a principle of organization. Things that are dead or have never been alive do not have a multiplicity of Stories. So stones never have souls. Plants and animals have their sets of Stories that apply in addition to physics and chemistry Stories, namely Stories of nutrition, growth, prey-seeking, mating and many others; and therefore they have can be said to have (appropriate kinds of) souls. And humans each have a soul to which all kinds of human Stories apply, especially those using human language, with which we have been so concerned in this book. But there are still no ghosts in their machines!

PART 4
THINKING IT OVER

Chapter 15

Epilogue

> I will not choose what many men desire
> Because I will not jump with common spirits
> And rank me with the barbarous multitude.
>
> W. Shakespeare, *The Merchant of Venice.*

We have travelled a long way since I suggested that we could understand Saltarella's human nature if we approached it by examining the cause of her attempt to break the high-jump record. Where has our intellectual journey taken us? Why *did* Saltarella jump?

She jumped because she was a living human athlete who wanted to win a competition. Her body was made of many parts, many cells, and they in turn of chemicals, atoms and subatomic particles, but her training had brought all their functioning together into an anatomical and physiological excellence by which she could be identified as a superb athlete. To put it shortly, she was alive in a unique way: and because there are so many Stories to be told about her we can say that she had a soul and that her soul, the way she existed, was unique. Her whole person, known by the name *Saltarella*, was dedicated to winning, and her being aware of that goal, her intention and desire to win, was the cause of her jumping. It was not that her self or her mind or her will commanded her body to do things: it was not that her body did something and dragged her along with it: *she*, the person watched by the crowd, jumped. She chose to jump at what seemed to her the optimal time and place. There was no compulsion to jump just then, and just there, and in that exact way; and even if in part the action was triggered by automatic habits of pattern perception that enabled her to recognise the optimal place and moment, she knew afterwards that it was just what she would have wanted to do. She was able to perceive consciously and unconsciously the details of the situation, and in addition her long practice activated brain and muscle activities of which she was not, perhaps, consciously aware. But she used the information processed by her senses, and the activation of her hormonal responses to prepare her muscles for the supreme effort. She did not sit inside her head and command her body like a pilot in an aircraft: she embodied her intentions, and became the movements of her limbs. She looked at the bar; she thought for a few moments and prepared herself for the effort. She willed herself (not her *self*!) to jump, and in doing so her brain sent commands to the muscles in a synchronised burst of activity that projected her body upwards towards the bar, transcending the stochastic quantal events of muscle biochemistry by her command of her body (not of her molecules) to synchronise the metabolic processes in the muscles. This was so even though she was not aware of the actions of her nerve cells or her muscle fibres. Why should she have been, since they were not separate from her as a person, but part of how she lived at that moment? She, Saltarella, the person seen as an individual athlete by the watching crowd, neither more nor less, jumped, as a person, with the final cause of winning, the efficient cause of her training, the material cause of her bodily metabolism, and the formal cause her being a living human, that is having a human soul. She was not a spirit, even if she as a person transcended in her biography a mere physical description.

Then Saltarella went home, fell in love, married, made new humans, "and lived happily ever after" - a human being until she died, and subsequently a stirring memory. All her Stories became a single History Story of her life. She was a person who had been a living human being, and hence a creature with a body made of chemicals, and a set of abilities that constituted her mind; responsible for her choices and therefore someone with a will; and an integrated living individual, a person, and for that reason describable as having a soul; aware of her intentions, surroundings and thoughts, describable by science and by philosophy and many other kinds of Stories.

She was indeed a human being.

> One does not receive wisdom; one must discover it oneself after a journey which no one can make for us, or spare us the trouble of, because it consists in its essence of a point of view on things.
>
> Marcel Proust. *A l'ombre des jeunes filles en fleurs. 1919.*

Appendix 1

Formulae for statistics.

Suppose we have a set of 10 measurements that make up our data. Call these $x_1, x_2, ..., x_{10}$ with values as shown below.

Number of data points = N = 10

Calculating the mean.

datum	x_1	x_2	x_3	x_4	x_5	x_6	x_7	x_8	x_9	x_{10}
value of datum	5	7	11	7	8	10	9	9	6	12

The mean, μ, is calculated by adding up all the values and dividing by the number of data points.

$$\mu = \Sigma(x)/N$$

$$= (5 + 7 + ... + 6 + 12)/10$$

$$= 84/10 = 8.4$$

Calculating the standard deviation, σ

Based on the same data, the standard deviation, σ is calculated as the square root of the variance of the values of the data. We calculate the deviation of each datum from the mean, square that number, sum the results, divide by the number of data points, and take the square root of the result.

	x_1	x_2	x_3	x_4	x_5	x_6	x_7	x_8	x_9	x_{10}
value of datum	5	7	11	7	8	10	9	9	6	12

Mean = 8.4

Deviation from mean   -3.4 -1.4 2.6 -1.4 -0.4 1.6 0.6 0.6 -2.4 3.6
Deviation^2 (D^2)   11.6 2.0 6.8 2.0 0.2 2.6 0.4 0.4 5.8 13.0

Sum of D^2 = 44.4

Mean of D^2 = 44.4/10 = 4.44

σ = √4.44 = 2.11

So if we collected more data, about 68% of the data would be expected to lie between 8.4 ± 2.11, that is between 6.3 and 10.5 . Only about 2% would lie above 12.6 .

Appendix 2
Bayesian Probability

$P(H|D) = P(D|H)*P(H)/P(D)$

where

H is a hypothesis, and D is the data.

P(H) is the prior probability of H: the probability that H is correct before the data D were seen.

P(D|H) is the conditional probability of seeing the data D given that the hypothesis H is true. P(D|H) is called the *likelihood*.

P(D) is the probability of getting the data, D.

P(H|D) is the posterior probability: the probability that the hypothesis is true, given the data and the previous state of belief about the hypothesis.

P(D) is the prior probability of witnessing the data D under all possible hypotheses.

Given any exhaustive set of mutually exclusive hypotheses H_i, we have:

$P(D) = \sum_i P(D, H_i) \sum_i P(D|H_i)P(H_i)$

Given any situation where we have an initial estimate of the probability of a hypothesis being true, and we then obtain some data about what has actually happened, we can, using these equations, calculate the impact of the data on our estimate of the probability of the hypothesis being true. Likewise, we can calculate the expected probability of obtaining the data if the hypothesis is true.

Bibliography

Ashby, W.R. 1956. *Introduction to Cybernetics.* London: Chapman and Hall.
Attenborough, D. 2013. *First Life.* BBC DVD.
Auden, W. H. 1969. *Oh tell me the truth about love.* In *Collected Shorter Poems, 1927-1957.* London. Faber and Faber.
Bach-Y-Rita. P. 1969. Vision substitution by tactile image projection, *Nature, 221*, 963-964.
Balzac, H. de, *La Comédie Humaine.* 1842.
Barber, T.X. 1969. *Hypnosis: a Scientific Approach.* New York. Van Nostrand.
Beishon, R. J. 1974. An analysis and simulation of an operator's behaviour in controlling continuous baking ovens. In Edwards, E. and Lees, F. (eds) *The Human Operator in Process Control.* London. Taylor and Francis.
Bernstein. J. 1998. *Against the Gods.* New York. John Wiley & Sons.
Bostrom, N. *Superintelligence: Paths, Dangers, Strategies.* Oxford Universty Press. 2014
Boswell, J. *Life of Johnson.* 1799.
Byron, Lord. *Don Juan.* 1819.
Capek, K. 1920. *R.U.R. (Rossum's Universal Robots).* London. Penguin Edition 2004.
Chalmers, D. 1998. *The Conscious Mind.* Oxford. Oxford University Press.
Chesterton, G.K. 1914. *The Flying Inn.* London. Methuen.
Chomsky, N. 1967. *A Review of B.F.Skinner's Verbal Behavior.* In L. Jacobovitz and M. Miron (eds). *Readings in the Psychology of Language.* New York. Prentice-Hall.
Coleridge, S. T. 1826. Quoted in R.Holmes, 1998. *Coleridge.* London. Harper Collins, p.550.
Conway, F. & Siegelman. 2005. *Dark Hero of the Information Age: in Search of Norbert Wiener.* New York. Basic Books.
Cooper, E. 1958. *That Uncertain Midnight.* London. Hodder Paperback.
Daily Telegraph. London. 19 June 2009. *Natural selection of moths.*

Darwin. C. 1859. *On the Origin of Species.* 2008. Oxford. Oxford World Classics.

Darwin, C. 1871. *Descent of Man.* 2004. London. Penguin.

Davis, R. Moray, N. Treisman, A. M. 1961. Imitative responses and the rate of gain of information. *Quarterly Journal of Experimental Psychology, 13.* 78-89.

Dawkins, R. 1988. *The Blind Watchmaker.* London. Penguin Books.

Dawkins, R. 1996. *Climbing Mount Improbable.* New York. W.W. Norton.

Dyson, F. 1979. *Disturbing the Universe.* New York. Harper and Row.

Eccles, J. 1953.*The Neurophysiology of Mind.* Oxford. Oxford University Press.

Edwards, E. and Lees, F. (eds) 1974. *The Human Operator in Process Control.* London. Taylor and Francis.

Feynman, R. 1985. *QED: the Strange Theory of Light and Matter.* Princeton. Princeton Science Library.

Feigenbaum, E. A. and Feldman, J. 1963. *Computers and Thought.* New York. McGraw Hill.

Geach, P. T. 1969. *God and the Soul.* London. Routledge and Kegan Paul.

Geach, P. T. and Anscombe, E. 1954. *Descartes: Selected Writings.* Edinburgh. Nelson.

Gerard, R. W. 1946. The Biological Basis of Imagination. *Scientific Monthly*, June. Page 487.

Golding.W. 1955. *The Inheritors.* London. Faber and Faber.

Gregory, R. L., Moray, N. and Ross, H. 1965. The curious eye of *Copilia. Nature*, 201, 116-1168.

Hawking, S. and Mladinow, L. 2003. *The Grand Design.* London. Bantam Books.

Hodges, A. 2012. *Turing, the Enigma.* Princeton. Princeton University Press.

Hoffmann. P. 2013. *Life's Ratchet.* New York. Basic Books.

Humphrey. N. 2008. *Seeing Red: a Study in Consciousness.* Cambridge. Harvard University Press.

Jackson, P. 1998. *Introduction to Expert Systems.* New York. Addison-Wesley.

Jones, E.H. 1942. *The Road to En-dor.* London. Bodley Head.

Kahneman, D. 2011. *Thinking Fast and Slow.* London. Penguin Books.
Kenny, A. 1973. *Wittgenstein.* Harmondsworth. Penguin Books.
Kenny, A. 1989. *The Metaphysics of Mind.* Oxford. Oxford University Press.
Kipling. R. 1893. *McAndrew's Hymn.* In *Collected Poems.* Ware, Hertfordshire. Wordsworth Editions, 1994.
Kipling, R. 1898. *The Day's Work.* New York. Doubleday McClure.
Klein, G. 2009. *Streetlights and Shadows: Searching for the Keys to Adaptive Decision Making.* Cambridge, MA. MIT Press
Kohler. I. 1964. *The Formation and Transformation of the Perceptual World.* New York. International University Press.
Kurtzweil, R. 2011. *The Singularity is Near.* New York. Penguin Books.
Laird, J. 2012. *The SOAR Architecture.* Cambridge. MIT Press.
Lessing, D. 1994. *The Making of the Representative for Planet 8.* London. Flamingo Press.
Libet, B.; Gleason, C.A.; Wright, E.W.; Pearl, D. K. 1983. Time of Conscious Intention to Act in Relation to Onset of Cerebral Activity (Readiness-Potential). *Brain* 106 (3): 623–42.
McCulloch, W. 1951. *Why the mind is in the head.* In L. Jeffress, ed., *Cerebral Mechanisms in Behavior.* The Hixon Symposium. New York. Wiley.
McCulloch, W. 1965. *Embodiments of Mind.* Cambridge. MIT Press.
Moray. N. 1963. *Cybernetics: Machines with Intelligence.* London. Burns Oates.
Moray, N. 1972. Visual mechanisms in the copepod *Copilia. Perception*, 1, 193-207.
Moray, N., and K. J. Connolly. 1963. A possible case of the genetic assimilation of behaviour. *Nature*, 199, 353-359.
Nagel, T. 1974. What is it like to be a bat? *The Philosophical Review*, Vol. 83, No. 4. 435-450.
Nagel, T. 2012. *New York Review of Books.* Vol. LIX (19), 40.
Oswald, I., Taylor, A.M. & Treisman, M. 1960. Discriminative responses to stimulation during human sleep. *Brain, 83*, 440-453.
Peacock, H. A. 1940. *Elementary Microtechnique.* London. Edward Arnold.

Penfield, W. 1975. *The Mystery of the Mind.* Princeton. Princeton University Press.
Penrose, R. 2004. *The Road to Reality.* Vintage Books.
Pinker, S. 2002. *The Blank Slate.* London. Penguin Books.
Pliny. *In* Harvard Loeb Classical Library, 1969. *Letters: Books VIII-IX.* Cambridge. Harvard University Press.
Rose, S. 1997. *Lifelines.* Oxford. Oxford University Press.
Ryle, G. 1949. *The Concept of Mind.* London. Hutchinson.
Sachs, M. 1988. *Einstein versus Bohr.* Lascelles Illinois. Open Court.
Sagan, C. 1973. *The Cosmic Connection: an Extraterrestrial Perspective.* Cambridge. Cambridge University Press.
Scruton. R. 2012. *Brain Drain.* The Spectator, 17 March.
Shannon, C., and Weaver, W. 1947. *The Mathematical Theory of Communication.* Urbana: University of Illinois Press.
Shu-Hsien Liao, 2005. Expert systems methodologies and applications. – a decade review from 1995 to 2005. *Expert Systems and Applications, 28(1),* 93-103.
Simon, H.A. 1996. *The Sciences of the Artificial.* Cambridge, MA. MIT Press.
Spencer, H. 1879. *Principles of Ethics.*
Strawson. P.F. 1959. *Individuals.* London. Methuen.
Thurber, J. 1950. *The Thirteen Clocks.* New York. Yearling.
Tolkein, J.R.R. 1954. *The Two Towers. Lord of the Rings.*Volume 2. New York. Houghton Mifflin.
Treisman, A. 1964. Verbal cues, language and meaning in selective attention. *American Journal of Psychology, 77,* 206-219.
Veltman, M.G. and Veltman, M. 2003. *Facts and Mysteries in Elementary Particle Physics.* Singapore. World Scientific Publishing.
Venter, C. 2007. *A Life Decoded.* London. Penguin Books.
Venter, C. 2011. *Life at the Speed of Light.* London. Little,Brown.
Virgil. *Georgics.* (Edited by P. Fallon. 2006. Oxford. Oxford University Press.)
Von Neumann, J. 1951. *The general and logical theory of automata.* In L. Jeffress, ed., *Cerebral Mechanisms in Behavior.* The Hixon Symposium. New York. Wiley.
Von Senden, M. 1960. *Space and Sight.* London. Methuen.
Watson, G. 2003. *Free Will.* Oxford. Oxford University Press.

Weiner. J. 1995. *The Beak of the Finch.* New York. Vintage Books.
Wiener, N. 1958. *Cybernetics: communication and control in the animal and the machine.* Cambridge, MA. MIT Press.
Williams, C. 1931. *The Place of the Lion.* London. Victor Gollancz.
Wyndham, J. 1951. *The Day of the Triffids.* Harmondsworth. Penguin Books.
Wittgenstein, L. 1968. *Philosophical Investigations.* Oxford. Blackwells.
Wollstonecraft. W. 2003. *Frankenstein.* London, Penguin Books.
Yates, D. 1922. *Berry and Co.* London. Ward Locke.

URLS
which also appear as footnotes or in the text.

Chapter 1.
About Stories.
http://www.40kbooks.com/?p=2176

Chapter 5.
About sending information from one brain to another.
http://telegraph 28/2/2013

Chapter 7.
About synthesising life.
http://www.wired.com/wiredscience/2010/05/scientists-create-first-self-replicating-synthetc-life/
About nerve cell activity
http://en.wikipedia.org/wiki/Action_potential
About structures in cells.
http://en.wikipedia.org/wiki/Flagellum

Chapter 8.
About evolutionary sequences

http://www.darwiniana.org/landtosea.htm

Chapter 9.
About the chemical formulae of molecules
http://glossary.periodni.com/glossary.php?en=sucrose
About brains of dinosaurs
http://poetry.poetryx.com/poems/8026/ .
About the structure of neurons
http://vv.carleton.ca/~neil/neural/neuron-a.html
About the functioning of nerve cells.
http://people.eku.edu/ritchisong/301notes2.htm.
About the chemistry of metabolism
http://www.youtube.com/watch?v=-ygpqVr7_xs&feature=related.

Chapter 11.
About prostheses and robots
http://www.bbc.co.uk/news/health-17936302.
http://www.nlm.nih.gov/medlineplus/news/fullstory_132301.html
http//gizmodo.com/5587600/rex-bionics-has-the-technology
http://www.technologyreview.com/news/522086/an-artificial-hand
    -with-real-feelings/
http://www.sciencedaily.com/releases/2011/05/110511111957.html
About Kurtzweil
http://bigthink.com/videos/ray-kurzweil-your-brain-in-the-cloud
About auditory prostheses
http://en.wikipedia.org/wiki/Cochlear_implant
About visual prostheses
http://www.bbc.co.uk/news/health-17936302.
About artificial problem solving
http://en.wikipedia.org/wiki/Logic_Theorist
About artificial intelligence
http://tp://act-r.psy.cmu.edu

Chapter 12.
About inverted visual world

http://books.google.fr/books?id=XNsDAAAAMBAJ&lpg=PA114
&ots=BEC_Ata71J&dq=inverting%20spectacles&pg
=PA114#v=onepage&q=inverting%20spectacles&f=false

Chapter 13
About social pressure and behaviour
http://en.wikipedia.org/wiki/Milgram_experiment.
About brain and free will
http://aspire.princeton.edu/news/archive/neurophilosophy.xml

Index

A

Actus hominis 239, 265, 267

Actus humanis 239, 265, 267

AI 186, 188, 192, 194, 197, 198, 200, 205, 206, 208

Aquinas 106, 219, 237, 248, 259, 261, 268, 271, 272, 274, 276

Aristotle 29, 31, 33, 106, 114, 132, 134, 135, 220, 233, 237, 241, 248, 259, 268, 270, 272, 273, 276

Artificial intelligence 6, 8, 13, 184, 186, 188, 294

Atomic Force Microscope 26, 36, 83, 154, 164, 276

B

Bacteria 8, 115, 117, 125, 126, 130, 136, 152, 159, 276

Bayes 90

Big Bang 127, 165

Biochemistry 4, 9, 16, 23, 34, 36, 70, 122, 128, 143, 150, 151, 154, 155, 158, 192, 242, 250, 272, 282

Biography 9, 17, 21, 27, 33, 36, 71, 83, 105, 113, 114, 118, 167, 188, 215, 239, 242, 251, 264, 266, 270, 272, 275, 282

Blindsight 215, 216

Bottom-up 154, 158

Brain 4, 6, 8, 13, 16, 22, 25, 27, 31, 35, 38, 39, 43, 45, 49, 51, 53, 55, 57, 67, 77, 102, 119, 120, 122, 160, 162, 171, 187, 188, 190, 192, 204, 206, 209, 210, 212, 214, 216, 218, 220, 222, 224, 226, 228, 230, 235, 240, 242, 247, 250, 253, 254, 256, 271, 277, 282, 291, 292, 294

C

Cartesian ego 107, 109, 220

Cause 9, 12, 14, 17, 21, 22, 25, 27, 29, 31, 33, 36, 37, 40, 49, 55, 59, 68, 70, 73, 79, 81, 88, 95, 107, 124, 125, 133, 135, 141, 144, 150, 152, 161, 170, 178, 215, 235, 236, 241, 248, 250, 252, 259, 260, 263, 268, 270, 272, 273, 281, 282

Cell ix, 4, 7, 8, 13, 17, 22, 25, 32, 36, 52, 57, 83, 116, 118, 120, 121, 124, 125, 127, 129, 132, 134, 135, 147, 150, 153, 156, 159, 160, 161, 164, 167, 169, 170, 189, 190, 194, 204, 214, 217, 218, 224, 228, 237, 238, 240, 248, 258, 262, 264, 269, 272, 274, 275, 278, 281, 282, 293, 294

Chalmers 231, 289

Chance 56, 76, 77, 79, 82, 85, 87, 90, 91, 93, 96, 126, 130, 131, 138, 139, 141, 144, 145, 170, 175, 177, 267

Chemistry 4, 12, 16, 23, 29, 33, 36, 46, 57, 128, 131, 133, 135, 153, 155, 156, 158, 162, 164, 166, 205, 241, 242, 247, 259, 276, 278, 294

Chlorophyll ix, 121, 156, 159

Clone 4, 150, 151, 153, 202

Cogito ergo sum 271

Complexity theory 67, 81, 128

Computer, computers 8, 97, 110, 113, 140, 141, 185, 187, 191, 192, 193, 194, 195, 197, 198, 199, 201, 202, 206, 208

Conditioning 72, 179, 243

Consciousness xiii, 4, 21, 36, 37, 39, 49, 53, 101, 102, 104, 106, 163, 209, 210, 215, 216, 218, 220, 222, 224, 226, 228, 230, 232, 259, 290

Cybernetics 5, 31, 34, 47, 186, 187, 273

D

Darwin, Charles 56, 64, 137, 139, 141, 150, 151, 290, 298

Darwin's Finches 150, 151

Dawkins, Richard 137, 143, 148, 242, 290

Death 4, 35, 46, 48, 79, 107, 114, 116, 121, 132, 133, 202, 258, 262, 268, 271

Dementia 221

Descartes 48, 49, 104, 106, 108, 110, 114, 188, 209, 219, 220, 231, 232, 259, 261, 262, 271, 274, 290

Diarese 110

DNA 8, 16, 57, 116, 120, 123, 126, 129, 133, 143, 156, 164, 167, 169, 170, 171, 177, 180, 182, 202, 206, 238, 242, 270, 272, 277

Dualism xiii, 48, 219, 220, 229, 231

E

Eccles 95, 247, 290

Efficient cause 31, 33, 34, 60, 96, 133, 272, 273, 282

Ego 51, 107, 108, 209, 210, 219, 220, 222, 227, 232, 241, 273

Einstein 11, 62, 66, 93, 95, 229, 292

Entropy 129, 134

Enzymes 8, 123

ESP 73, 75, 77

Everyday stories 4, 11, 37, 45, 48, 74, 135, 168, 171, 182, 188, 209, 257, 258, 268, 272, 277

Evolution ix, 6, 13, 14, 93, 136, 137, 140, 141, 143, 145, 148, 149, 152, 153, 179, 233, 275

Experiment 43, 50, 56, 59, 61, 64, 65, 68, 69, 73, 75, 77, 79, 83, 87, 89, 91, 94, 95, 118, 130, 142, 143, 210, 213, 216, 227, 230, 232, 238, 246, 253, 254, 256, 295

Explanation 16, 21, 22, 27, 29, 34, 37, 40, 51, 59, 61, 63, 65, 70, 71, 73, 75, 77, 91, 94, 96, 104, 132, 163, 217, 218, 231, 270

Extrasensory perception 73

F

Fable 15, 29, 31, 40, 41, 50, 53, 68, 84, 87, 94, 101, 145, 177, 182, 218, 225, 226, 229, 251, 276

Feynman 66, 229, 230, 290

Final cause 31, 33, 73, 133, 272, 282

fMRI 7, 36, 43, 45, 68, 210, 218, 222, 224, 230, 234, 241, 277

Formal cause 31, 33, 272, 273, 282

Fossils 145, 149

Frankenstein 116, 163, 293

Freedom 237, 239, 240, 249, 250, 252

Free will xiii, 4, 8, 12, 29, 38, 95, 105, 236, 239, 240, 242, 245, 247, 248, 253, 256, 262, 292, 295

Fundamental Words 5, 8, 35, 37, 39, 42, 45, 55, 69, 73, 97, 101, 102, 105, 106, 110, 112, 167, 208, 257

G

Gene 36, 63, 150, 168, 179, 183, 242

Genetic algorithm 140

Ghost xiii, 38, 39, 46, 47, 49, 51, 53, 91, 109, 141, 163, 200, 203, 207, 209, 210, 219, 224, 232, 234, 253, 258, 263, 269, 271, 277

Ghost in the Machine 39, 48, 163, 207, 253, 258, 269

GIM 48, 49, 135, 200, 201, 218, 224, 227, 230, 233, 234, 237, 253, 258, 263

God vi, 11, 14, 27, 36, 54, 55, 68, 79, 116, 136, 141, 152, 172, 262, 268, 273, 289, 290

God spot 54, 55, 68

Gravity 53, 61, 65, 95, 230

Greek 11, 17, 23, 31, 75, 80, 106, 110, 160, 173, 183, 233, 258, 260, 268

Gunfighter paradox 255

H

Haemoglobin ix, 121, 156, 159

Hard Problem, the 211, 212

Heredity 4, 167, 168, 170, 179

Heritability 6, 178

High-jump 22, 25, 31, 33, 281

Hoffmann 23, 26, 83, 93, 119, 125, 129, 131, 164, 166, 276, 290

Human development 262

Hypnosis 244, 246, 251, 289

I

Instinct 72, 73, 167

Intelligence xiii, 4, 6, 8, 13, 35, 37, 43, 51, 79, 101, 106, 167, 168, 170, 172, 177, 178, 181, 182, 184, 186, 188, 198, 200, 202, 204, 259, 291, 294

IQ ix, 6, 172, 173, 174, 176, 178, 181, 200

K

Kahneman 43, 201, 216, 239, 266, 291

Kenny xv, 35, 45, 47, 49, 51, 69, 226, 230, 232, 257, 261, 291

Krebs cycle ix, 120, 164, 166

L

Language 10, 12, 17, 26, 33, 35, 39, 41, 43, 45, 48, 49, 51, 53, 55, 57, 69, 71, 83, 101, 102, 104, 106, 108, 110, 112, 114, 116, 132, 153, 169, 171, 172, 174, 176, 179, 188, 196, 207, 215, 216, 218, 220, 222, 225, 228, 230, 232, 234, 240, 242, 252, 256, 258, 264, 268, 275, 277, 289, 292

Latin 9, 106, 108, 110, 117, 184, 233, 237, 239, 258, 260, 266

Life xv, 4, 6, 8, 10, 13, 14, 16, 21, 22, 26, 30, 33, 36, 37, 43, 52, 53,

55, 58, 79, 83, 93, 96, 101, 102, 106, 110, 114, 116, 118, 119, 121, 124, 125, 128, 130, 132, 134, 135, 153, 160, 162, 164, 166, 169, 171, 176, 179, 184, 191, 198, 201, 202, 206, 210, 212, 218, 224, 226, 228, 233, 234, 238, 240, 242, 246, 248, 252, 257, 258, 261, 262, 264, 268, 270, 272, 275, 277, 282, 289, 290, 292

Lloyd Morgan's Canon 71, 73

M

Material cause 32, 33, 60, 70, 73, 133, 270, 272, 273, 282

Materialism 46, 53, 112

Materialist xiii, 38, 39, 44, 53, 242

Mendel 63, 229

Mental arithmetic 47, 206

Mental events 16, 21, 26, 36, 53, 105, 163, 218, 228, 241

Middle Ages 29, 56, 118, 259

Mind xiii, 4, 6, 9, 28, 34, 35, 37, 39, 44, 45, 47, 49, 53, 64, 71, 73, 75, 79, 95, 101, 102, 104, 106, 108, 110, 171, 180, 188, 192, 200, 206, 209, 210, 212, 215, 217, 220, 223, 225, 226, 231, 234, 236, 240, 242, 247, 257, 259, 261, 281, 282, 289, 290, 292

Mind and brain 215

Mind-body problem 45, 105

Mitochondria 8, 119, 122, 126, 128, 164, 166, 248

Model 26, 59, 61, 63, 65, 68, 69, 77, 79, 87, 90, 127, 149, 161, 168, 178, 202, 212, 215, 226, 228, 230

Molecules 8, 22, 24, 26, 30, 33, 36, 51, 119, 121, 123, 125, 128, 129, 132, 154, 155, 156, 157, 164, 189, 191, 203, 241, 248, 270, 277, 278, 282, 294

Muscle 4, 9, 22, 24, 25, 27, 32, 34, 49, 53, 70, 89, 94, 104, 146, 147, 160, 161, 167, 189, 190, 235, 237, 238, 250, 255, 264, 277, 282

Muscle fibre 22, 24, 25, 34, 282

N

Nagel vi, 223, 224, 291

Names 5, 31, 44, 101, 102, 104, 107, 216, 227, 269

Natural selection 139, 141, 147, 152, 178, 289

Nature and Nurture 167, 168, 172

Necessity 80, 93, 130

Nerve cell 4, 7, 13, 26, 36, 52, 57, 119, 122, 124, 160, 189, 190, 204, 214, 217, 218, 225, 237, 238, 240, 264, 282, 293, 294

Neural nets 191

Neuron ix, xi, 7, 26, 36, 52, 77, 160, 161, 167, 189, 190, 193, 194, 196, 208, 215, 218, 294

Neuroscience xiii, 5, 8, 39, 68, 87, 212, 217, 231, 235, 253, 255, 256, 262

Newton 14, 61, 65, 95, 231

Normal distribution ix, 84, 180

Nouns 101, 102, 104, 211, 232

Nucleic acid 8, 123, 124, 132

O

Ockham's Razor 70, 71, 73

Origin of species 137, 141, 290

Other minds 221

P

Pasteur 118

Person xiii, 5, 11, 14, 21, 22, 34, 36, 38, 39, 45, 51, 54, 55, 72, 74, 82, 83, 90, 108, 110, 115, 121, 135, 151, 153, 158, 162, 165, 167, 168, 170, 172, 176, 178, 179, 190, 199, 200, 202, 204, 210, 216, 218, 226, 229, 230, 232, 234, 238, 240, 242, 244, 246, 248, 250, 254, 258, 262, 266, 269, 271, 273, 277, 278, 281, 282

Philosophical Story 27, 51, 104, 135

Philosophy xv, 4, 6, 12, 14, 16, 28, 29, 34, 35, 39, 42, 43, 45, 50, 51, 55, 67, 69, 73, 87, 97, 106, 112, 118, 131, 163, 235, 237, 239, 256, 258, 277, 282

Physics 4, 8, 12, 16, 23, 27, 29, 32, 33, 36, 39, 46, 48, 53, 56, 57, 65, 67, 69, 76, 80, 82, 87, 93, 95, 122, 125, 128, 131, 132, 136, 153, 155, 158, 161, 207, 214, 229, 230, 232, 241, 242, 247, 250, 259, 270, 276, 278, 292

Physiology 4, 9, 16, 167, 224, 239, 253

Pitts-McCulloch neurons 193, 194, 197

Plato 11, 44, 103, 106, 241, 259, 268

Pliny 108, 292

Predictability 67, 95, 247, 249

Prediction 8, 29, 60, 61, 63, 65, 67, 69, 78, 82, 95, 242

Probability ix, 5, 6, 9, 24, 25, 42, 64, 68, 69, 76, 77, 79, 81, 84, 85, 87,

89, 91, 93, 95, 97, 130, 131, 137, 149, 154, 173, 177, 243, 287

Prosthetics 184, 187

Proteins 22, 25, 120, 122, 124, 127, 130, 158, 164, 191

Psychology xiv, xvi, 8, 12, 16, 26, 53, 57, 67, 69, 83, 96, 163, 216, 227, 289, 290, 292

Q

Quantum Theory 8, 66, 94, 124, 154, 206, 229, 231

R

Race 6, 38, 43, 174, 175, 177, 178, 181, 182

Racial 43, 172, 175, 176, 177

Randomness 80, 96

Random numbers 96, 97

Reductionism 53

Religion 3, 13, 14, 54, 55, 68, 79, 118, 152, 258

Responsibility 236, 240, 247, 248, 251, 252, 256

Robotics 13, 187, 194, 205, 207, 208

Robots 34, 119, 128, 166, 185, 187, 198, 201, 202, 276, 294

S

Saltarella 9, 10, 21, 22, 24, 25, 28, 31, 33, 36, 37, 39, 46, 49, 51, 79, 94, 96, 125, 132, 160, 167, 170, 179, 180, 236, 239, 240, 243, 250, 270, 277, 281, 282

Science vi, xiii, xv, 3, 4, 6, 8, 12, 14, 16, 21, 22, 29, 31, 34, 35, 38, 39, 41, 43, 45, 53, 55, 57, 59, 61, 63, 66, 67, 69, 73, 76, 77, 79, 82, 87, 94, 95, 97, 101, 106, 112, 118, 127, 131, 136, 137, 144, 149, 163, 168, 181, 187, 192, 201, 207, 208, 229, 231, 235, 242, 256, 258, 268, 273, 282, 290, 292

Scientific Stories 12, 15, 16, 23, 28, 31, 34, 36, 39, 45, 56, 73, 78

Self vi, 4, 10, 12, 15, 37, 45, 49, 51, 59, 61, 64, 67, 69, 96, 116, 123, 127, 128, 129, 137, 143, 191, 200, 203, 209, 210, 232, 234, 267, 277, 281, 282, 293

Soul xiii, 4, 6, 8, 28, 33, 35, 37, 46, 48, 49, 51, 53, 102, 106, 110, 114, 118, 132, 134, 135, 203, 219, 235, 236, 257, 258, 260, 262, 264, 266, 268, 270, 271, 273, 275, 277, 281, 282, 290

Species ix, 23, 71, 123, 125, 128, 132, 137, 141, 143, 145, 147, 149, 152, 156, 157, 182, 215, 223, 224, 234, 267, 290

Spiritualism 46, 48

Statistics 68, 83, 86, 96, 172, 175, 177, 180, 285

Stem cells 132, 167, 264, 274, 275

Stories 1, 4, 6, 10, 12, 15, 16, 21, 22, 25, 27, 31, 34, 36, 37, 39, 45, 48, 49, 51, 56, 67, 73, 78, 94, 111, 112, 135, 136, 158, 168, 171, 182, 188, 209, 256, 258, 268, 272, 278, 281, 282, 293

T

Teleology 34, 273

Telepathy 73, 89

Theory vi, 6, 8, 46, 50, 57, 59, 61, 63, 65, 67, 69, 71, 77, 79, 81, 87, 94, 95, 115, 124, 128, 137, 144, 154, 203, 206, 229, 230, 290, 292

Thinking xv, 1, 5, 17, 19, 28, 35, 42, 43, 55, 72, 74, 99, 107, 108, 110, 117, 188, 190, 199, 200, 206, 210, 216, 219, 220, 222, 224, 228, 230, 234, 239, 240, 248, 251, 260, 266, 273, 274, 279, 291

Top-down 159

Truth table xi, 106, 196

Turing 192, 194, 197, 198, 200, 203, 204, 206, 290

Turing machines 192, 194, 203, 204, 207

Twins 150, 169, 177

W

Weiner 144, 150, 152, 186, 209, 273, 293

Wiener 5, 31, 34, 187, 289, 293

Wittgenstein 28, 35, 44, 51, 106, 224, 257, 258, 291, 293

Z

Zygote 132, 150, 159, 160, 162, 165, 264, 266, 267

A thorough review of a wide range of issues spanning such concepts as "soul" and "free will" and the role science plays with its insistence on replicability and its ability to undermine those concepts. *O.F.G. Sitwell, Professor Emeritus, University of Alberta.*

In *"Science, Cells, and Souls,"* Neville Moray, scientist, philosopher, and psychologist lays out a distillation of a lifetime's wisdom. He attacks life's most difficult and daunting questions that bedevil all human beings. Reminiscent of Descartes and Russell, Moray illuminates the crucial issue as to what connotes a meaningful question. He shows us how to distinguish the boundary conditions in this most problematic of all enquires. In so doing, he acts as a modern-day Virgil, guiding us surely and safely across the yawning chasms of incipient ignorance to lead us to the broad uplands of a comprehensive enlightenment. But Moray is no arid empiricist clinging dogmatically to the orthodoxies of standard scientific doxology. He provides us with ideas to wrestle with, to chew on, to mull over – it is a simply brilliant exposition that offers wisdom and an intellectual and emotional empathy which characterizes the man himself. It is the culmination of a lifetime's research. You can choose to ignore it but if you purchase it you can revolutionize your view of what constitutes what we are pleased to call reality. *Peter Hancock, Provost's Distinguished Research Professor and Pegasus Professor, University of Central Florida.*